VEHICULAR 2017

The Sixth International Conference on Advances in Vehicular Systems, Technologies and Applications

July 23 - 27, 2017

Nice, France

VEHICULAR 2017 Editors

Markus Ullmann, BSI & H-BRS, Germany

Khalil El-Khatib, University of Ontario, Canada

Calin Vladeanu, University "Politehnica" of Bucharest (UPB), Romania

Manabu Tsukada, University of Tokyo, Japan

Printed from e-media with permission by:

Curran Associates, Inc.
57 Morehouse Lane
Red Hook, NY 12571

Some format issues inherent in the e-media version may also appear in this print version.

Printed by Curran Associates, Inc. (2017)

International Academy, Research, and Industry Association (IARIA)
412 Derby Way
Wilmington, DE 19810

Phone: (408) 893-6407
Fax: (408) 527-6351

petre@iaria.org

Additional copies of this publication are available from:

Curran Associates, Inc.
57 Morehouse Lane
Red Hook, NY 12571 USA
Phone: 845-758-0400
Fax: 845-758-2633
Email: curran@proceedings.com
Web: www.proceedings.com

Table of Contents

VEHICULAR 2017

Foreword

The Sixth International Conference on Advances in Vehicular Systems, Technologies and Applications (VEHICULAR 2017), held between July 23 - 27, 2017 - Nice, France, continued the inaugural event considering the state-of-the-art technologies for information dissemination in vehicle-to-vehicle and vehicle-to-infrastructure and focusing on advances in vehicular systems, technologies and applications.

Mobility brought new dimensions to communication and networking systems, making possible new applications and services in vehicular systems. Wireless networking and communication between vehicles and with infrastructure have specific characteristics from other conventional wireless networking systems and applications (rapidly-changing topology, specific road direction of vehicle movements, etc.). These led to specific constraints and optimizations techniques; for example, power efficiency is not as important for vehicle communications as it is for traditional ad hoc networking. Additionally, vehicle applications demand strict communications performance requirements that are not present in conventional wireless networks. Services can range from time-critical safety services, traffic management, to infotainment and local advertising services. They are introducing critical and subliminal information. Subliminally delivered information, unobtrusive techniques for driver's state detection, and mitigation or regulation interfaces enlarge the spectrum of challenges in vehicular systems.

We take here the opportunity to warmly thank all the members of the VEHICULAR 2017 Technical Program Committee, as well as the numerous reviewers. The creation of such a high quality conference program would not have been possible without their involvement. We also kindly thank all the authors who dedicated much of their time and efforts to contribute to VEHICULAR 2017. We truly believe that, thanks to all these efforts, the final conference program consisted of top quality contributions.

Also, this event could not have been a reality without the support of many individuals, organizations, and sponsors. We are grateful to the members of the VEHICULAR 2017 organizing committee for their help in handling the logistics and for their work to make this professional meeting a success.

We hope that VEHICULAR 2017 was a successful international forum for the exchange of ideas and results between academia and industry and for the promotion of progress in the field of vehicular systems, technologies and applications.

We are convinced that the participants found the event useful and communications very open. We also hope that Nice provided a pleasant environment during the conference and everyone saved some time for exploring this beautiful city.

VEHICULAR 2017 Chairs:

VEHICULAR Steering Committee

Markus Ullmann, Federal Office for Information Security / University of Applied Sciences Bonn-Rhine-Sieg, Germany
Carlos T. Calafate, Universitat Politècnica de València
Éric Renault, Institut Mines-Télécom | Télécom SudParis, France
Samy El-Tawab, James Madison University, USA
Khalil El-Khatib, University of Ontario Institute of Technology - Oshawa, Canada
Manabu Tsukada, University of Tokyo, Japan

VEHICULAR Industry/Research Advisory Committee

Clément Zinoune, Renault, France
Michelle Wetterwald, HeNetBot, France
Yi Ding, US Army RDECOM-TARDEC, USA
William Whyte, Security Innovation, USA

VEHICULAR 2017

Committee

VEHICULAR Steering Committee

Markus Ullmann, Federal Office for Information Security / University of Applied Sciences Bonn-Rhine-Sieg, Germany
Carlos T. Calafate, Universitat Politècnica de València
Éric Renault, Institut Mines-Télécom | Télécom SudParis, France
Samy El-Tawab, James Madison University, USA
Khalil El-Khatib, University of Ontario Institute of Technology - Oshawa, Canada
Manabu Tsukada, University of Tokyo, Japan

VEHICULAR Industry/Research Advisory Committee

Clément Zinoune, Renault, France
Michelle Wetterwald, HeNetBot, France
Yi Ding, US Army RDECOM-TARDEC, USA
William Whyte, Security Innovation, USA

VEHICULAR 2017 Technical Program Committee

Ali Abedi, University of Maine, USA
Eric J. Addeo, DeVry University, USA
Sufyan T. Faraj Al-Janabi, University of Anbar, Ramadi, Iraq
Michele Albano, CISTER | INESC/TEC | Polytechnic Institute of Porto, Portugal
Alexandre Armand, Renault, France
Dalila B. Megherbi, University of Massachusetts, USA
Eduard Babulak, Fort Hays State University, USA
Andrea Baiocchi, University of Roma "Sapienza", Italy
Stylianos Basagiannis, United Technologies Research Centre, USA
Marcel Baunach, Graz University of Technology | Institute for Technical Informatics, Austria
Luca Bedogni, University of Bologna, Italy
Rahim (Ray) Benekohal, University of Illinois, USA
Chafika Benzaid, University of Sciences and Technology Houari Boumediene (USTHB), Algeria
Luis Bernardo, Universidade NOVA de Lisboa, Portugal
Neila Bhouri, IFSTTAR / COSYS / GRETTIA, France
Abdelmadjid Bouabdallah, Université de Technologie de Compiègne, France
Christos Bouras, University of Patras | Computer Technology Institute & Press «Diophantus», Greece
Salah Bourennane, Ecole Centrale Marseille, France
Lin Cai, Illinois institute of Technology, USA
Carlos T. Calafate, Universitat Politècnica de València
Roberto Caldelli, CNIT - National Interuniversity Consortium for Telecommunications | MICC - Media Integration and Communication Center, Universita' degli Studi Firenze, Italy

Maria Calderon, University Carlos III of Madrid, Spain
Juan-Carlos Cano, Universidad Politécnica de Valencia, Spain
Rodrigo Capobianco Guido, São Paulo State University at São José do Rio Preto, Brazil
Juan Carlos Ruiz, Universitat Politecnica de Valencia, Spain
Sandra Céspedes, Universidad de Chile, Santiago, Chile
Amitava Chatterjee, Jadavpur University, Kolkata, India
Claude Chaudet, Webster University Geneva, Switzerland
Olfa Chebbi, Higher Institute of Management of Tunis, Tunisia
Jiajia Chen, KTH Royal Institute of Technology, Sweden
Mu-Song Chen, Da-Yeh University, Taiwan
Dong Ho Cho, School of Electrical Eng., KAIST, Korea
Gihwan Cho, Chonbuk University, Korea
Woong Cho, Jungwon University, South Korea
Theofilos Chrysikos, University of Patras, Greece
Yousef-Awwad Daraghmi, Palestine Technical University-Kadoorie, Palestine
Klaus David, University of Kassel, Germany
David de Andrés, Universitat Politècnica de València, Spain
Carl James Debono, University of Malta, Malta
Sergio Di Martino, University of Naples Federico II, Italy
Omar Dib, Technological Research Institute IRT SystemX, France
Yi Ding, US Army RDECOM-TARDEC, USA
Sanjay Dorle, G. H. Raisoni College of Engineering, Nagpur, India
Mariagrazia Dotoli, Politecnico di Bari, Italy
Ghaïs El Zein, Institut d'Electronique et de Télécommunications de Rennes (IETR) | Institut National des Sciences Appliquées (INSA), France
Safwan El Assad, University of Nantes, France
Mohamed El Kamili, University Sidi Mohammed Ben Abdellah of Fez, Morocco
Khalil El-Khatib, University of Ontario Institute of Technology - Oshawa, Canada
Samy El-Tawab, James Madison University, USA
Marcos Fagundes Caetano, University of Brasília, Brazil
Camille Fayollas, ICS-IRIT | University Toulouse 3 Paul Sabatier, France
Gustavo Fernandez, Austrian Institute of Technology, Austria
Gianluigi Ferrari, University of Parma, Italy
Attilio Fiandrotti, Politecnico di Torino, Italy
Miguel Franklin de Castro, Federal University of Ceará, Brazil
Małgorzata Gajewska, Gdansk University of Technology, Poland
Sławomir Gajewski, Gdansk University of Technology, Poland
David Gallegos, Applus+ IDIADA, Spain
Peter Gaspar, MTA SZTAKI, Hungary
Puya Ghazizadeh, Millersville University, USA
Thanassis Giannetsos, University of Surrey, UK
Apostolos Gkamas, University Ecclesiastical Academy of Vella of Ioannina, Greece
Javier Gozalvez, Universidad Miguel Hernandez de Elche, Spain
Luigi Alfredo Grieco, Politecnico di Bari, Italy
Stefanos Gritzalis, University of the Aegean, Greece
Yu Gu, Hefei University of Technology, China
Mesut Güneş, Institute for Intelligent Cooperating Systems | Otto-von-Guericke-University Magdeburg, Germany

Biao Han, National University of Defense Technology, Changsha, China
Petr Hanáček, Brno University of Technology, Czech Republic
Hong Hande, National University of Singapore, Singapore
Morteza Hashemi, Ohio State University, USA
Hiroyuki Hatano, Utsunomiya University, Japan
Ivan W. H. Ho, The Hong Kong Polytechnic University, Hong Kong
Javier Ibanez-Guzman, Renault S.A., France
Khalil ibrahimi, Ibntofail University, Kenitra, Morocco
Hocine Imine, IFSTTAR/LEPSIS, France
Mohammad Reza Jabbarpour, Islamic Azad University, Tehran-North Branch, Iran
Terje Jensen, Telenor, Norway
Han-You Jeong, Pusan National University, Korea
Yiming Ji, University of South Carolina Beaufort, USA
Magnus Jonsson, Halmstad University, Sweden
Yasin Kabalci, Omer Halisdemir University,Turkey
M. A. S. Kamal, Monash University Malaysia, Malaysia
Rossi Kamal, University of Asia Pacific, Bangladesh
Frank Kargl, Ulm University, Germany
Sokratis K. Katsikas, University of Piraeus, Greece
Abdelmajid Khelil, Landshut University, Germany
Wolfgang Kiess, DOCOMO Euro-Labs, Germany
Wooseong Kim, Gachon University, Korea
Xiangjie Kong, Dalian University of Technology, China
Jerzy Konorski, Gdansk University of Technology, Poland
Dimitrios Koukopoulos, University of Patras, Greece
Zdzislaw Kowalczuk, Gdansk University of Technology, Poland
Milan Krbálek, Czech Technical University in Prague, Czech Republic
Fedor Lehocki, Slovak University of Technology in Bratislava, Slovak Republic
Christian Lehsing, Harvard Medical School, USA & Technical University of Munich, Germany
Pierre Leone, University of Geneva, Switzerland
Marco Listanti, University of Roma "La Sapienza", Italy
Lianggui Liu, Zhejiang Sci-Tech University, China
Miao Liu, IBM Research, USA
Yali (Tracy) Liu, AT&T labs, Inc., USA
Seng Loke, Deakin University, Melbourne, Australia
Xuanwen Luo, Sandvik Mining, USA
Barbara M. Masini, CNR - IEIIT, Italy
Leandros Maglaras, De Montfort University, UK
Abdelhamid Mammeri, University of Ottawa, Canada
Zoubir Mammeri, IRIT - Paul Sabatier University, France
Suzanne McIntosh, Cloudera, Inc. | New York University, USA
Natarajan Meghanathan, Jackson State University, USA
Rashid Mehmood, King Abdul Aziz University, Jeddah, Saudi Arabia
João Mendes-Moreira, LIAAD-INESC TEC | University of Porto, Portugal
Lyudmila Mihaylova, University of Sheffield, UK
Vicente Milanes, RENAULT SAS, France
Saeid Nahavandi, Institute for Intelligent Systems Research and Innovation (IISRI) - Deakin University, Australia

Markus Ullmann, Federal Office for Information Security / University of Applied Sciences Bonn-Rhine-Sieg, Germany
Rens W. van der Heijden, Ulm University, Germany
Adriaan J. van Wijngaarden, Bell Laboratories | Nokia, USA
John Vardakas, Iquadrat Informatica, Barcelona, Spain
Massimo Villari, Universita' di Messina, Italy
Ljubo Vlacic, Griffith School of Engineering, Australia
Lingfeng Wang, University of Wisconsin-Milwaukee, USA
You-Chiun Wang, National Sun Yat-sen University, Taiwan
Shuangqing Wei, Louisiana State University, USA
Andre Weimerskirch, Lear Corporation, USA
Michelle Wetterwald, HeNetBot, France
William Whyte, Security Innovation, USA
Safwan Wshah, PARC - a Xerox Company, USA
Fan Wu, Tuskegee University, USA
Pei Xiao, Institute for Communication Systems (ICS) | University of Surrey, UK
Ramin Yahyapour, Gesellschaft für wissenschaftliche Datenverarbeitung mbH Göttingen (GWDG), Germany
Fan Ye, Stony Brook University, USA
Paul Yoo, Bournemouth University, UK
Shingchern D. You, National Taipei University of Technology, Taiwan
Chuan Yue, Colorado School of Mines, USA
S.M. Salim Zabir, National Institute of Technology | Tsuruoka College, Japan
David Zage, Intel Corporation, USA
Sherali Zeadally, University of Kentucky, USA
Degan Zhang, Tianjin University of Technology, China
Liang Zhang, Magna Electronics, Brampton Ontario, Canada
Clément Zinoune, Renault, France

Copyright Information

Quality of Service Assessment in Connected Vehicles

Asiya Khan and Sanjay Sharma
School of Engineering,
Plymouth University, Plymouth, PL4 8AA UK
Email: asiya.khan;sanjay.sharma@plymouth.ac.uk

Cindy Goh[1] and Cheng Leong Lim[2]
[1]School of Engineering, Glasgow University,
Glasgow G12 8LT UK
[2]NXP Semiconductors, Singapore 138628
Email:cindy.goh@glasgow.ac.uk;
chengleong.lim@nxp.com

Abstract— **In recent years, there has been a huge interest in Machine-to-Machine connectivity under the umbrella of Internet of Things (IoT). With the UK Government looking to trial autonomous (driverless) cars this year, connected vehicles will play a key part in improving and managing existing road safety and congestion, leading to a new generation of intelligent transport systems. This is also well aligned to the current initiatives by the automotive industry to improve the driver's experience on-board. However, the wireless channels most suitable for this application have not been standardized. In this paper, we review the wireless channels suitable for vehicle-2-vehicle (V2V) and Vehicle–to-x (V2x) connectivity. We further present preliminary analysis on the factors that impact the Quality of Service (QoS) of connected vehicles. We use the open access GEMV^2 data to carry out Analysis of Variance (ANOVA) and Principal Component Analysis (PCA) on the link quality and found that both line of sight and non line of sight has a significant impact on the link quality. The work presented here will help in the development of connected vehicle network (CVN) prediction model and control for V2V and V2x connectivity. It will further contribute towards unfolding and testing key research questions in the context of connected vehicles which may otherwise be overlooked.**

Keywords-QoS; V2V; V2x; ANOVA; PCA; CVN.

I. INTRODUCTION

Improving road safety and traffic management on existing infrastructure has huge societal and economic impacts. While users' perception of vehicular experience is fast evolving, existing transport infrastructure is still playing catch up. There is a disadvantage in using proprietary wireless standard as it limits customizability due to the lack of open source movement. Therefore, standardization of wireless channels is a first step to addressing this challenge as it allows for interoperability and user multiplier effect. The availability and presence of wireless communications and connectivity in vehicles is shaping customers' on-board experience. In the same manner that mobile phones have evolved in the last ten years, vehicles are evolving with some form of vehicle-2-vehicle (V2V) connectivity. According to a market research report, connected vehicles will be worth $46.69 billion by 2020 [1]. Connected vehicles are defined as a set of moving networked computer systems with dozens of electronic control units (ECUs), hundreds of sensors and million lines of code [2]. Research investigating the suitability of wireless channels is a significant starting point to them becoming a reality in the near future [3][4]. The benefits of V2V connectivity especially in areas of collision avoidance and congestion management are huge, V2V is becoming a reality and automobile industry is currently working towards standardization.

A number of developed countries are trialling autonomous cars on the roads in the near future. Google's cars have driven 1.2 million miles in USA, with Germany, China and the UK [5], also looking to open trials. Connected vehicles will play a key part in traffic management of autonomous cars. Within the next five years there will be some form of autonomous driving on UK roads. It is therefore important to investigate the best wireless channels in this application so as to fully understand the challenges and be able to address them effectively. This will also help towards the modelling of wireless channels for connected vehicles.

A number of researchers have presented findings both on technique [6] and a network model [7]. Petri nets are proposed in [6] for such time critical distributed communication and control systems. GEMV^2, a geometry-based V2V channel model has been presented in [7], which measures link quality by factoring outlines of vehicles, buildings, and foliage to distinguish between the three types of links; the links are Line of Sight (LOS), Non-LOS due to vehicles and Non-LOS (NLOS) due to static objects. In addition, the link quality is calculated with the large-scale signal variation deterministically and the small scale-signal variation stochastically based on the number and size of surrounding objects. GEMV^2 is freely available to be used by researchers.

The objective of this paper is to identify and present those challenges and opportunities associated with Quality of Service (QoS) in connected vehicles and to identify the modelling direction for Connected Vehicle Network (CVN) by conducting Analysis of Variance (ANOVA) and Principal Component Analysis (PCA) on the factors that impact link quality. Here, we define CVN as the network between V2V and V2x and where the position and /velocity of the vehicle is predicted from the previous vehicle/x. The 'x' in V2x represents vehicle/infrastructure/roadside sensors/anything else deemed suitable. The vision for CVN is that each vehicle on the road will be able to communicate with other vehicles and this set of data and communication will support a new generation of active safety applications and systems [8]. Wireless technologies and their potential challenges in providing vehicle-to-x connectivity are presented in [4]. An overview of applications and associated requirements of vehicular networks are presented in [9]. Internet mobility in vehicular scenarios along with their challenges is presented in [10]. With ever increasing connectivity and a vision that migrates towards smart cities, security issues and the challenges such as propriety networks, inter-operability between networks, etc. therein are immense. Work in [11] presents some of the security challenges in vehicular ad hoc networks (VANET), whereas [12] focuses on the four working groups on scientific foundations of vehicular networking and presents their findings. Connected Vehicle Network is modelled using a black-box approach that comprises of vehicles with wireless V2V communication using link length estimator to identify the number of vehicles in the network [13], whereas [14] presented modelling of future state of a vehicle in a platoon based on preceding vehicle position and velocity.

In this paper, we use the data from $GEMV^2$ to carry out ANOVA and PCA. Doing so, helps us to better understand the QoS relationship between the link quality and the factors that impact it. We chose four factors that impact link quality as LOS, NLOS, number of neighbours per vehicle and the neigh-thresh per vehicle. The parameters are described in Section III.

The work presented in this paper differs from the ones listed above since it provides an in depth analysis on the various wireless channels available for connected vehicles based on our QoS assessment of the $GEMV^2$ data. Therefore, the contributions of the paper are two-fold:

- to present an overview of wireless channel requirements in connected vehicles.
- to present ANOVA and PCA on $GEMV^2$ data to understand the impact of line of sight, non line of sight, neighbours and neigh-thresh per vehicle on link quality. This enables us to present our modelling directions for CVN.

The rest of this paper is organized as follows. Section II gives an overview of the connected vehicle channel requirements. Section III describes the QoS assessment on $GEMV^2$ data, whereas Section IV discusses the research directions for CVN modelling. Conclusions and future work is presented in Section V.

II. CONNECTED VEHICLES CHANNEL REQUIREMENTS: AN OVERVIEW

The concept diagram of connected vehicles is presented in Figure 1, which illustrates V2V and V2x connectivity using various access networks which is in turn connected to the core network. The concept behind Figure 1 is that connected vehicles will be able to communicate with each other and with an intelligent transport system (ITS) using different wireless channels such as Wi-Fi, 4G/LTE, etc. QoS in such application will be critical as vehicles come out of one network into the other especially at handover points. Connected vehicles are the building blocks of emerging Internet of Vehicles (IoV) and Network of Things (NoT) [15], which is defined on five primitives as sensors, aggregator, communication channel, external utility and decision trigger. All vehicles or 'x' will have sensors connected that will be able to transmit/receive 'useful' information. This information is converted by an aggregator, defined as a mathematical function implemented in software that transforms raw data into some 'useful' meaning. This is underpinned by the communication channel, e.g., WiFi, 4G, etc. The external utility can be a software/hardware and will execute processes into the overall workflow of NoT. Finally, the decision trigger creates the final result needed to satisfy the requirements of NoT.

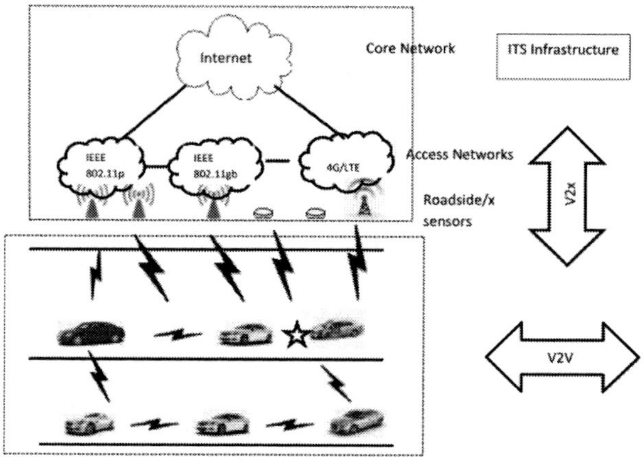

Figure 1. V2V and V2x concept diagram

The ITS reference architecture from [16] has been adapted and is presented in Figure 2. It is a protocol stack inspired from the Open Systems Interconnection (OSI) model and defines three layers as 'access', which will support the wireless access networks/wireless channels, a network & transport layer which supports the routing protocols, data transfer, etc. Above it sits the facilities layer which will support the application/information. Here, we define the position/velocity of the vehicle in this layer. The layers of application, management and security run across both horizontally and vertically and provides cross layer commands and information.

A number of applications ranging from infotainment, for example, media downloading to traffic safety applications, such as driving assistance co-operative awareness impose diverse requirements on supporting vehicular networking technologies. There will be a huge emphasis on inter-networking between the different standards in order to achieve seamless communications. In addition, there are different requirements for inter-vehicle (V2V or V2x) and intra-vehicles networks. Intra-vehicle is defined as all the ECUs within the vehicle communicating to the driver and includes infotainment. Hence, all the wireless channels described in this section may play a role in the connected vehicle application. Therefore, this section provides an overview on the wireless channels available and the connectivity challenges required in a V2V or V2x communication type.

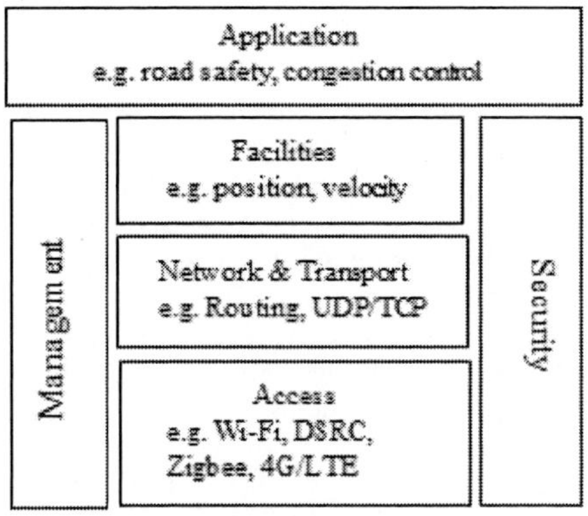

Figure 2. ITS reference traffic structure (adapted from [16])

A. DSRC/Wave

Dedicated short-range communications with wireless access in vehicular environments (DSRC/WAVE) as defined by IEEE 802.11p and IEEE1609 (higher layer standard based on IEEE 802.11p) is a key enabling wireless technology for both V2V and V2R communications. DSRC works in 5.9GHz band with a bandwidth of 75MHz in the US and 30MHz in Europe and an approximate range of 1000m. It is designed for both one way and two way communication. DSRC are not compatible in Europe, Japan and US. Currently, DSRC is the default broadcast communication protocol used. Some limitation of DSRC includes its dedicated spectrum in supporting V2V communication type [17] and lack of QoS support. Key application for DSRC is roadside sensors which transmit information about hazardous conditions, road surface and distance between vehicles and anti-collision information.

B. Zigbee

Zigbee is based on IEEE 802.15.4 specification intended for wireless personal area network applications with low power and cost. Zigbee also has applications in V2R connectivity where the moving vehicle exchanges information with the roadside sensors [18]. The Zigbee enabled roadside sensors then updates traffic status to an intelligent control system seamlessly. It also has application in intra-vehicle networking where a small wireless sensor network is established between the sensors.

C. Visible Light Communication (VLC)

The use of visible light communication (VLC) for V2R communication is proposed in [19]. VLC is defined by IEEE 802.15.7 standard and can support data rate up to 96Mb/s through fast modulation of LED light sources [19]. It is an emerging area of research given the possibility of augmenting existing infrastructure such as traffic lights. However, one key limitation of VLC is any poor weather conditions such as rain and fog could ultimately degrade its communication reliability.

D. Wi-Fi

Wi-Fi standards are based on IEEE 802.11 series, mainly using the 2.4/5GHz band. A number of automobile manufacturers are building new cars with in-built Wi-Fi capability, providing infotainment applications. V2V connectivity could also foster the integration of bicycles and pedestrians into the networks [10] using Wi-Fi. This has a huge potential in improving road safety and reducing the number of accidents as a result of blind spots.

TABLE I. SUMMARY OF WIRELESS CHANNELS FOR V2V AND V2X COMMUNICATION TYPES

Wireless Channels	Advantages	Disadvantages
DSRC/WAVE	Default broadcast network currently used	Limited coverage, (~1000m), QoS not supported
Zigbee	Mesh network, scalable, no need for centralized control	Low and limited data rate, not mature security, limited coverage (10-100m)
VLC	Infrastructure already there, 1-2000m range	Early stages/cost of conversion
Wi-Fi	Widely implemented, 35m indoor and 115m outdoor	Interoperability with other protocols
4G/LTE	Existing infrastructure, several Km range	Interoperability with other protocols

E. 4G/LTE

Long-Term Evolution (LTE) is a standard for high speed communications for mobile phones and data terminals. The standard is developed by 3GPP. The key advantage of LTE-connected cars [4] is having cars connecting directly to the Internet through existing 4G-LTE cellular network. Work in [20] presents a hybrid scheme that can achieve seamless IP communication over mobile Internet access.

F. Summary of CVN Channel Requirements

Table I summarize the various wireless channels, their standard requirements and potential advantages and disadvantages for V2V and V2x. The current industry trends are choosing DSRC and 4G/LTE as the best way to offer connectivity between cars. Many critical applications are linked to safety applications, e.g., air bag control, automatic braking, etc. Inter-operability between these networking standards will be an important milestone. Work presented in [21] concludes that DSRC configuration choice has an impact on safety messages successfully transmitted. In addition, as suggested in [22][23], an upper limit on information provided to the vehicle may be necessary to prevent overloading drivers with information. This poses additional requirements and challenges towards the standardization of wireless channels for vehicle communication. Depending on the communication type e.g., V2V or V2x, all of the wireless channels presented in Table I will be relevant and the CVN modelling has to take that into account.

III. QOS ASSESSMENT IN CONNECTED VEHICLES

This section presents the QoS assessment using Analysis of Variance (ANOVA) and Principal Component Analysis

(PCA) on GEMV^2 data for V2V and V2I. This will help us in understanding the interaction between the four parameters chosen and their impact on the link quality and lay the foundation in establishing the modelling direction for CVN.

A. GEMV^2

GEMV^2 (Geometry-based Efficient Propagation Model for V2V communication) [7] data is freely available and is implemented in MATLAB. GEMV^2 measures large-scale variation calculated deterministically and small-scale signal variation stochastically based on the number and size of the surrounding objects. Both the signal variation is measured in decibels. We use the GEMV^2 data of large-scale and small-scale signal variation under the influence of four different conditions - they are LOS, NLOS, the number of neighbouring vehicles and the neigh-thresh per vehicle. The data is available for both V2V and V2I. The communication channel is IEEE802.11p.

Table II. ANOVA Results for Main and Interaction Effects for V2V & V2I Data

Source	Sum of Squares	Degree of freedom	Mean Squares	F-statistics	p-value
V2V					
LOS	23646.7	38	622.283	5.37	0
NLOS	18100	39	464.102	4	0
Neighbours	6377.9	41	155.558	1.34	0.0828
Neigh-thresh	189.3	4	47.321	0.41	0.8028
LOS*NLOS	66.9	1	66.9451	17.73	0.0007
LOS*Neighbours	141	3	47.0094	12.45	0.002
NLOS*Neighbours	24.6	1	24.64.13	6.52	0.0212
Neighbours*Neigh-thresh	34.4	1	34.3572	9.1	0.008
V2I					
LOS	340.4	7	48.625	2.5	0.0181
NLOS	12669.6	20	633.479	32.63	0
Neighbours	60.3	7	8.614	0.44	0.8733
Neigh-thresh	1248.2	6	208.027	10.72	0
LOS*NLOS	69.4	2	34.71	0.52	0.6244
LOS*Neighbours	0	2	0.017	0	0.9998
NLOS*Neighbours	33	14	2.357	0.04	1
Neighbours*Neigh-Thresh	0	1	0	0	0.9995

LOS links have an unobstructed path between communicating vehicles, whereas NLOS is obstructed by vehicles and buildings. Neighbours is defined as the number of transmitting vehicles in the network and neigh-thresh is defined as the number of neighbouring vehicles whose received power was above the threshold.

B. ANOVA on GEMV^2 Data

ANOVA was carried out on the GEMV^2 dataset. ANOVA is chosen as it enables us to understand the interaction between the four parameters on link quality. Table II presents the results. ANOVA was carried out on large-scale signal variation only for both V2V and V2I as the interaction between parameters was found to be not as significant for small-scale variation.

Tables II shows the results of the ANOVA. The p-value is derived from the cumulative distribution function of F [24] and a small p-value indicates that the link quality is significantly influenced by the corresponding parameter. Between V2V communications, both LOS and NLOS have significant impact on the link quality, whereas between V2I communications, NLOS is slightly more significant than LOS and the Neigh-thresh have a higher impact on link quality. However, for V2V, all four parameters have small p-values indicating that they all in varying degree are significant. However, it is interesting to note that, in V2I, the number of neighbours per vehicle is not that significant. For V2V, the combined interaction between LOS and NLOS and NLOS and Neighbours is most significant. Whereas, for V2I, the combined interactions are less significant compared to the individual. To better understand the interactions, PCA investigation is carried out.

C. PCA on GEMV^2 Data

PCA was chosen as it reduces the dimensionality of the data while retaining as much information as possible. PCA involves calculating the eigenvalues and their corresponding eigenvectors of the covariance or correlation matrix. Covariance matrix is used where the same data has the same set of variables and correlation matrix is used in the case where data has a different set of variables. In this paper, covariance matrix was used because of the same dataset.

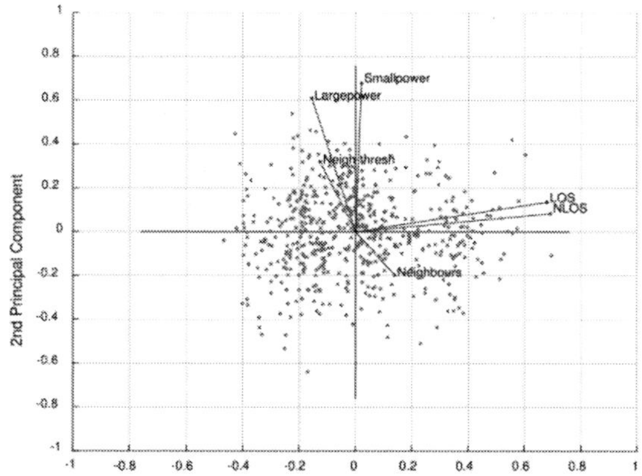

Figure 4a. PCA results for V2V

Figures 4a and 4b show the PCA results for V2V and V2I respectively. In addition to the four factors, both large-scale (Largepower) and small-scale (Smallpower) signal variation is used. The horizontal axis represents the first principal component and the vertical axis the second. Each of the parameters is represented by a vector. There are six components in Figures 4a and the first three components account for more than 90% of the variance. Figure 4a shows the first principal component contributes largely to LOS and NLOS.

Figure 4b shows the PCA results for V2I. Similar to Figure 4a, Figure 4b the first three components account for over 80% of the variability. Points on the edge of the plot have the lowest scores for the first principal component.

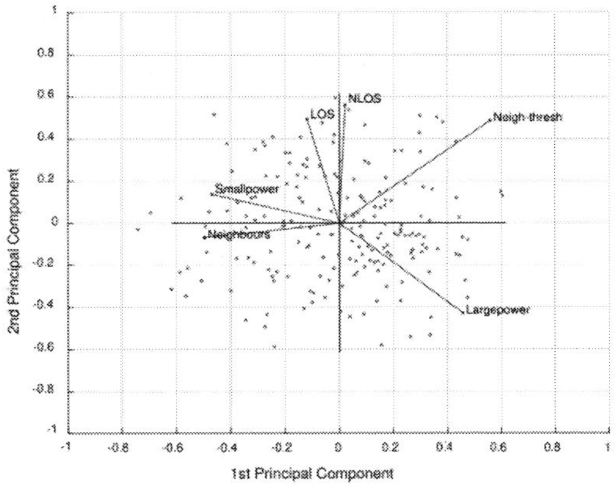

Figure 4b PCA results for V2I

IV. RESEARCH DIRECTIONS FOR CVN MODELLING

QoS assessment will enable us to choose parameters in modelling the CVN. Our small-scale QoS assessment highlighted some of the research challenges and hence potential opportunities for further work are as follows:

(i) Overcoming QoS issues in connected vehicles is fundamental to the successful deployment of V2x connectivity. The QoS can be affected by networking parameters such as bandwidth, delay and latency. In addition, parameters such as the distance between vehicles, road-side sensors and the speed of the vehicle all play a part towards the QoS of the V2x network thus integrating connected vehicles into IoT ecosystems [25]. QoS will be further divided between V2x service reliability for safety related applications where parameters such as time-sensitivity during message transfer, guarantee of message delivery, etc. are highest priorities. While, QoS of on-board applications e.g., infotainment will be lower in priority.

(ii) We also identified that the needs for trade-off between the amount of intelligence sitting with the vehicle for intra-vehicle connectivity and to that controlled remotely via an intelligent control system. Different wireless channels will be suitable for inter-vehicle vs intra-vehicle connectivity. For example, on-board sensors that can sense a motorbike/bicycle within the blind spot of the driver can greatly improve road safety and reduce accidents.

(iii) Prediction of CVN will be based on information centric network paradigm which is independent of location. The CVN will be predicted from the preceding state of the vehicle based on position/velocity.

Figure 5 shows an overview of real time sharing of sensor information between vehicles via cloud or V2Cloud (via DSRC or LTE-direct). It shows various scenarios of connectivity to clouds, ITS, other vehicles, pedestrians, etc. The Society of Automotive Engineers (SAE) has established communication standards for DSRC for connected vehicles (SAE J2735) [26]. This is the first step towards standardizing the CVN communication protocols as most vehicle manufacturers in the near future will be building cars with in-built Wi-Fi capability. An immediate application would be to reduce traffic congestion by relaying an accident/roadworks/incident to re-route traffic thus reducing the overall traffic congestion.

Therefore, the main research questions that emerge from existing literature and our QoS assessment on connected vehicles are:

- What are the QoS issues in V2x networks e.g., what impact(s) would weather conditions have on the QoS?
- What will the network standards for CVN be?
- Will there be different network protocols within CVN e.g., for V2V, V2R, V2I, etc. and what are the interoperability challenges here?
- How does CVN benefit autonomous car evolution?
- How to enhance security in CVN?
- What form(s) of connected vehicles will we see in the next five years on the roads of most developed economies?

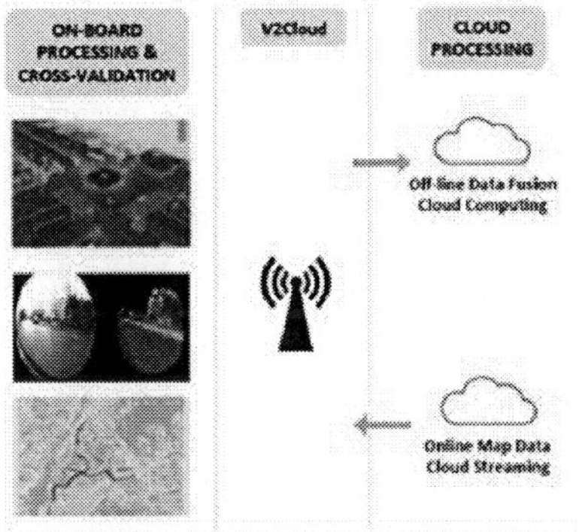

Figure 5. An overview of real-time sensor sharing V2Cloud [2]

V. CONCLUSIONS

This paper presents QoS assessment from ANOVA and PCA on the link quality of connected vehicles. We used data from $GEMV^2$. Our analysis shows that for V2V number of transmitting vehicles in the network (neighbours) has a bigger impact than in V2I on link quality. However, parameters of LOS and NLOS are significant in both types (V2V and V2I). This will help us determine the direction of modelling of the link network for CVN. It further addresses QoS challenges in connected vehicles and presents an overview of the various wireless channels and their applications in connected vehicles scenarios. The key issues identified will help lay the foundation for future research directions in this area. Some of the challenges that need to be addressed by wireless channels in connected vehicles are weather conditions and their impacts, for example how low visibility and extreme weather conditions can impact on the QoS of the connected vehicle. In addition, cameras and ultra-sonic sensors are limited to low distance. The overall reliability of the sensor data within connected vehicle communication is critical. As suggested in [3], for safety management, sensors that can detect fatigue levels of the driver by monitoring various bodily conditions can also be added. The first commercial vehicles to have onboard units installed are expected in summer 2017 from Cadillac [27].

The data information and filters necessary are also investigated, e.g., what is critical, necessary, add-on to process in the vehicle and what data to send/receive to/from the data centre. The challenge is to maintain the QoS of the real-time communication protocol and how to ensure data integrity of the process. With autonomous driving being trialled this year in the UK, what role will connected vehicles play? These are some of the imminent research questions highlighted from our research. Future direction of our research will aim to address the points raised in this paper and focus on modelling the CVN with some form of control.

ACKNOWLEDGMENT

The work reported here is supported in part by the Santander Seed-Corn Research Grant.

REFERENCES

[1] "Connected Car Market" by Connectivity Form Factor (Embedded, Tethered, Integrated), Connectivity Technology (LTE, 3G, HSPA+, Wi-Fi, Bluetooth), Application (Navigation, Telematics, Infotainment) and Geography - Trends & Forecast to 2014 – 2020.

[2] IEEE webinar on connected vehicles Part I and II, Dec. 2015.

[3] M. Faezipour, M. Nourani, A. Saeed and S. Addepalli, "Progress and challenges in intelligent vehicle area networks", Communications of the ACM, vol. 55, no.2, pp. 90-100, Feb. 2012.

[4] N. Lu, N. Cheng, N. Zhang, X. Shen and J. W. Mark, "Connected vehicles: solutions and challenges, IEEE Internet of Things Journal", vol. 1, no. 4, pp 289-299, August 2014.

[5] The pathway to driverless cars, summary report and action plan. Department of Transport UK, Feb. 2015.

[6] P. Diekhake, T. Kurczveil and U. Becker, "Modelling and simulation of distributed time critical communication and control systems in vehicles or vehicle networks", IFAC Conference Proceedings, 49-3, pp 303-308, 2016.

[7] M. Boban, J. Barros and O. Tonguz, "Geometry-based vehicle-to-vehicle channel modeling for large-scale simulation", IEEE Transactions on Vehicular Technology, vol. 63, Issue 9, pp. 4146-4164, 2014.

[8] S. Narla, "The evolution of connected vehicle technology: from smart drivers to smart cars to .... Self-driving cars", ITE Journal, July 2013.

[9] G. Karagiannis, O. Altintas, E. Ekici, G. Heijenk, B. Jarupan, K. Lin, and T. Weil, "Vehicular Networking: A Survey and Tutorial on Requirements, Architectures, Challenges, Standards and Solutions", IEEE Communications Surveys & Tutorials, Vol. 13, No. 4, Fourth Quarter 2011.

[10] S. Céspedes, X. Shen and C. Lazo, "IP Mobility Management for Vehicular Communication Networks: Challenges and Solutions", IEEE Communications Magazine May 2011, pp. 187–194.

[11] G. Samara, W. A.H. Al-Salihy and R. Sures, "Security Issues and Challenges of Vehicular AdHoc Networks (VANET)", , 4th International Conference on New Trends in Information Science and Service Science (NISS), 11-13 May 2010, Gyeongju, South Korea.

[12] F. Dressler, H. Hartenstein O. Altintas, and O. K. Tonguz, "Inter-Vehicle Communication: Quo Vadis", IEEE Communications Magazine, June 2014.

[13] L. Zhang and G. Orosz, "Black-box modelling of connected vehicle networks", In American Control Conference (ACC), Boston Mariott, Copley Place, July 6-8, Boston, MA, 2016.

[14] R. Pandita and D. Caveney, "Preceding vehicle state prediction", IEEE Intelligent Vehicles Symposium (IV), June 23-26, Gold Coast, Australia, 2013.

[15] J. Voas, "Network of 'Things', NIST Special publication", 800-183, 2016. http://nvlpubs.nist.gov/nistpubs/SpecialPublications/NIST.SP.800-183.pdf.[retrieved: May, 2017]

[16] Intelligent Transport Systems (ITS); Communications Architecture, September 2010. ETSI EN 302 665 V1.1.1 (2010-09).

[17] M. I. Hassan, H. L. Vu, and T. Sakurai, "Performance analysis of the IEEE 802.11 MAC protocol for DSRC safety applications," IEEE Trans. Veh.Technol., vol. 60, no. 8, pp. 3882–3896, Oct. 2011.

[18] J. Qiongqiong, Y. Zhou, J. Wang, Y. Wang and X. Wang "A lightweight cross-layer cooperative testbed for evaluation of connected vehicles", , 11th International Conference on Mobile Ad-hoc and Sensor Networks (MSN), 16-18 Dec. 2015, Shenzhen, China.

[19] S-H. Yu, O. Shih, and H-M. Tsai, "Smart Automotive Lighting for Vehicle Safety", IEEE Communications Magazine, Dec. 2013.

[20] S. Cespedes and X. Shen, "On achieving seamless IP communications in heterogenous vehicular networks", IEEE Transactions on Intelligent Transportation Systems, vol. 216, No. 6, Dec 2015.

[21] J. B. Kenney, S. Barve and V. Rai, "Comparing communication performance of DSRC OBEs from multiple suppliers", 19th ITS World Congress, 22-26 Oct., Austria, 2012.

[22] J.-M. Girard, N. Tricot, K. Younsi, and J.-C. Popieul, "When does the driver workload reaches its limits?" in Proc. IEEE Intell. Transp. Syst. Conf. (ITSC'06), Toronto, ON, Canada, Sep. 2006, pp. 578–583.

[23] C. Wu and Y. Liu, "Queuing network modeling of driver workload and performance," IEEE Trans. Intell. Transp. Syst., vol. 8, no. 3, pp. 528–537, Sep. 2007.

[24] G. W. Snedecor and W. G. Cochran, "Statistical Methods", 8th ed. Ames:Iowa State Univ. Press, 1989.

[25] S. K. Datta, R. P. F. Costa, J. Ham and C. Bonnet, "Integrating connected vehicles in Internet of Things ecosystems: challenges and solutions", IEEE 17th International Symposium of A World of Wireless, Mobile and Multimedia Networks (WoWMoM), 21-24 June 2016.

[26] Society of Automotive Engineers, "Dedicated Short Range Communications (DSRC) Message Set Dictionary", J2735 Ground Vehicle Standard, J2735, 2016.

[27] General Motors. Cadillac to Introduce Advanced "Intelligent and Connected" Vehicle Technologies on Select 2017 Models. [Online] [retrieved: May, 2017].

A Review of Network Models for Internet of Vehicles

Asiya Khan and Sanjay Sharma
School of Engineering
University of Plymouth, Drake Circus
Plymouth PL4 8AA
e-mail: asiya.khan;sanjay.sharma@plymouth.ac.uk

Cindy Goh[1] and Cheng Leong Lim[2]
[1]School of Engineering, Glasgow University,
Glasgow G12 8LT UK
[2]NXP Semiconductors, Singapore 138628
e-mail: cindy.goh@glasgow.ac.uk;
chengleong.lim@nxp.com

***Abstract*—Connected vehicles are the building blocks of the emerging Internet of Vehicles (IoV) under the umbrella of Internet of Things (IoT) and more recently, Network of Things (NoT). This paper applies the NoT concept to IoV networks and presents a review of the network models for IoV highlighting the research challenges and solutions.**

***Keywords-* IoV; V2V; V2x; NoT; VANET.**

I. Introduction

Automobiles are currently undergoing a revolution just like mobile phones did ten years ago. Future automobile will be expected to communicate with other cars – vehicle-2-vehicle (V2V), with infrastructure/roadside sensors/pedestrians/cyclists/anything else (V2x). The vision of smart cities includes connected vehicles amongst many other things. It is envisaged that 25 million 'things' will be connected by the year 2020. In one year, globally approximately 1.3 million lives are lost and 7.4 million injured in road accidents and 90 billion hours lost due to traffic delays [1]. Therefore to improve road safety and traffic congestion, connected vehicles offer a very promising solution. They also contribute towards the roadmap of fully autonomous driving becoming a reality. The emerging Internet of Vehicles (IoV) is offering the platform to provide real time exchange of information to realize the opportunity of improving road safety and congestion. It has huge applications in autonomous car revolution, intelligent transportation system and smart city.

The key technologies for IoV are presented in [2], guidelines and basic principles of IoV are presented in [3], whereas [4] focuses on the solutions and challenges for connected vehicles. As IoV revolution takes off, the conventional Vehicular Ad hoc Networks (VANETs) are changing into IoV as VANET [5] turns the connected vehicle into a wireless router or mobile node enabling vehicles to connect to each other creating a wireless network between them. An in-depth tutorial on vehicular networking is presented in [6], whereas [7] presents the challenges of integrating connected vehicles to Internet of Things (IoT) and [8] presents a vehicular cloud for IoV applications. There are a number of challenges within the IoV network based on the priority of data exchange messages. For example, priority has to be given to safety critical messages, whereas on-board messages related to infotainment will be lower on that scale. Work presented in [9] proposes an abstract network model for IoV based on individual and swarm activities. Petri-nets have been used recently in vehicular authentication [10], modelling and control of vehicular networks [11] and traffic signal analysis in [12]. Work presented in [13] models vehicular networks using spatio-temporal locality and information-centric networks (ICN) are presented in [14] to model the connected networks. Recently, the concept of Network of Things (NoT) with IoT has been presented in [15]. In this paper, we apply the concept of NoT to the emerging IoV as presented by NIST [15] and review the connected network models presented in literature identifying the challenges and solutions.

The rest of this paper is organized as follows. Section II presents the IoV concept, whereas an overview of the network models is presented in Section III. Section IV presents the research challenges and solutions for network models in IoV. Section V concludes the paper and presents future direction of our work.

II. NoT Applied to IoV

This section presents an overview of IoV, summarizes the wireless channels standards used in V2V and V2x and applies the NoT concept to IoV.

A. *IoV Concept*

IoV integrates three networks – an inter-vehicle network, an intra-vehicle network and vehicular mobile Internet. Therefore, IoV integrates these three networks and is defined as "a large-scale distributed system for wireless communication and information exchange between V2x according to agreed communication protocols and data standards" [9].

An inter-vehicle network is defined as network communication generated by the vehicle-borne computer, control system, on-board sensors, or passengers that is disseminated in the proximity to other vehicles. An intra-vehicle network is a wireless network between sensors inside a vehicle. A review of intra-vehicle networks is presented in [16]. The number of sensors is forecasted to reach 200 per vehicle by 2020 [1]. Electronic Control units (ECUs) are built in the vehicle which communicates to other ECUs and sensors wirelessly. Connected vehicles require Onboard units (OBUs) to broadcast messages through

VANET. Work presented in [17] shows that vehicle density has an impact on VANET network metrics. The OBUs will contain data such as vehicle location, current time, direction, speed, traffic volume remarks, acceleration, deceleration. The communication between two or more OBUs is the V2V communication and that between an OBU and RSU (Roadside Sensor Unit) is V2I. Recently, the Society of Automotive Engineers (SAE) has established communication standards for connected vehicles (SAE J2725) [18] under Dedicated Short Range Communication (DSRC) for V2V and V2x. Table I summarizes the various wireless communication channels with their requirements for typical communication types, e.g. V2V or V2I, etc.

TABLE I. SUMMARY OF WIRELESS CHANNEL STANDARDS

Channel	Frequency band	Bandwidth	Data rate	Range	Communication Type
DSRC/ WAVE	5.9GHz	75MHz in USA 30MHz in Europe	27 Mbps (max)	1000m	V2V & V2I/R/x
Zigbee	2.4GHz/868 MHz (Europe)	2 MHz	20kbps-250kbps	10-100m	V2I/V2R
VLC [19]	400 and 800THz	~390THz	10Kbps (signals) 500Mbps (LEDs)	1000-2000m	V2I/R
Wi-Fi	2.4/5 GHz	20/40 MHz	54Mbps-600Mbps	35m (indoor), 115m (outdoor)	V2V & V2x
4G/LTE	700/800/900/1800/2600 MHz in Europe Supported by IMT and ITU	20 MHz	300 Mbps peak download rates, 75 Mpbs upload rates	Worldwide – limited to cellular coverage zones	V2V & V2x

The sensors in an intra-vehicle communication are stationary so a simple star network topology is sufficient. In this communication type, the wireless protocols that support smaller distance is recommended, e.g. Bluetooth, Zigbee, Radio-Frequency Identification (RFID), etc. A vehicular mobile network is the cloud-based mobile network that sits above both inter-vehicle and intra-vehicle networks. Figure 1 gives an overview of the V2x connectivity using various wireless protocols. The concept behind Figure 1 is that connected vehicles will be able to communicate with each other and with an intelligent transport system (ITS) using different wireless channels such as Wi-Fi, 4G/LTE, etc. and integrated with various sensors. Quality of Service (QoS) in such application will be critical as vehicles come out of one network into the other especially at handover points.

There have been a number of researchers who have exploited both Wi-Fi [20]-[22] and DSRC/WAVE [23]-[26] in V2V and V2x communication. Line of Sight was achieved in [20][21] but when restricted by obstacles (no line of sight) then communication was affected. Work in [22] recommend 10MHz for V2V and V2x. Using WAVE [23], the environmental effects of antenna height, traffic and electromagnetic wave propagation had a severe impact on performance. The WAVE (IEEE802.11p) draft proposal is presented in [24] and in [25][26] authors confirm the viability of WAVE and IEEE802.11 in vehicular communication.

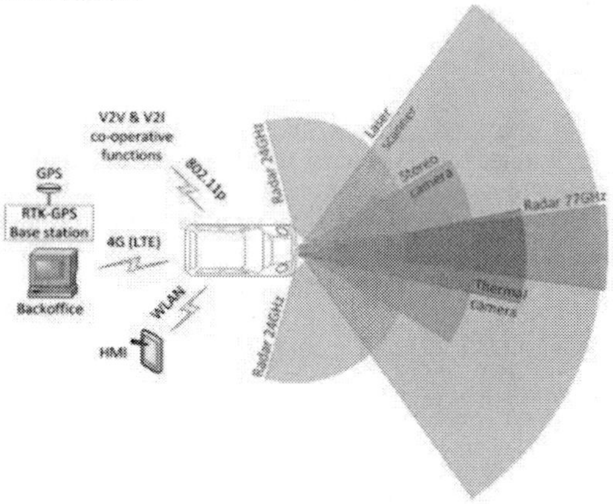

Figure 1. Future trends in connected vehicles [27]

While researchers have presented work on communication protocols, future network models of IoV needs to be compatible with all wireless protocols depending on the communication type e.g. V2V or V2x.

B. NoT applied to IoV

The concept diagram of connected vehicles under IoV is presented in Figure 2, which illustrates V2V and V2x connectivity using various access networks, which is in turn connected to the core network. The data is exchanged between Intelligent Transport Systems (ITS) and the vehicles. We link Figure 2 with NoT as presented by NIST [15].

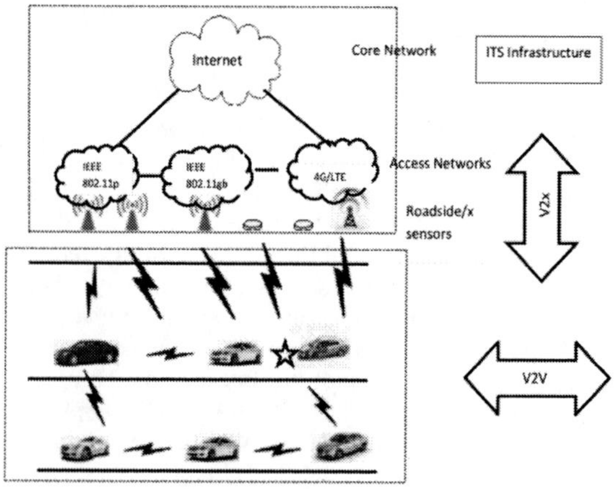

Figure. 2 V2V and V2x concept diagram under IoV

The NoT [15] defines five primitives as Sensors, Aggregator, Communication Channel, External Utility and Decision Trigger. All vehicles will have sensors connected that will be able to transmit/receive 'useful' information. This information is converted by an Aggregator, defined as a mathematical function implemented in software that transforms raw data into some 'useful' meaning. Both Sensor and Aggregator are shown as Roadside sensors in Figure 2. This is underpinned by the communication channel e.g. WiFi, 4G, etc. Again, Figure 2 shows the wireless channels such as Wi-Fi/4G etc. between V2V and V2x. The External Utility can be a software/hardware and will execute processes into the overall workflow of NoT. Finally, the Decision Trigger creates the final result needed to satisfy the requirements of NoT. The External Utility and Decision Trigger is combined together and presented within ITS in Figure 2.

TABLE II. IoV PRIMITIVES

NIST Primitives	Proposed Primitives	Feature
Sensor	Sensing Technologies	Wireless and wired, sensors, RFID,
Aggregate		
Communication Channel	Communication Channel	DSRC/Wave, Zigbee, Bluetooth, Wi-Fi, 4G/LTE
External utilities	Data Processing	Data created by connected vehicles, and how it is processed
Decision Trigger		

Based on these NoT primitives [15], we present three primitives. We combine the primitives of Sensor and Aggregator as just Sensing Technologies, Communication Channel and again combine External Utility (eUtility) and Decision Trigger as one and call it Data Processing as shown in Table II. In Table II, feature describes the potential features for each primitive.

III. IoV Network Models – An overview

This section presents an overview of the four network models presented in literature.

A. Petri-Net

Petri-nets combine a well-defined mathematical theory with a graphical representation of the dynamic behaviour of systems. Precise modelling and analysis of system behaviour is allowed by the theoretic aspect of Petri nets, whereas, the graphical representation of Petri nets enable visualization of the modelled system state changes [28].

Controller Area Network (CAN) communication bus has been modelled using petri nets in [10] and used for timing analysis. CAN [29] is generally used to transmit control traffic between ECUs within the vehicle (intra-vehicle) and is very popular in the automotive application as a communication bus for event-triggered communication. Petri nets offer huge potential for distributed communication which will be key in V2x communication type. For instance, petri-nets cope smoothly with defining and implementing complicated requests in VANET using tokens and measure, i.e. the limits of vehicle numbers of RSU groups.

Figure 3 shows the structure of a distributed control system which includes 2 vehicles, 6 devices, 10 programs and 12 functions with 10 data sets.

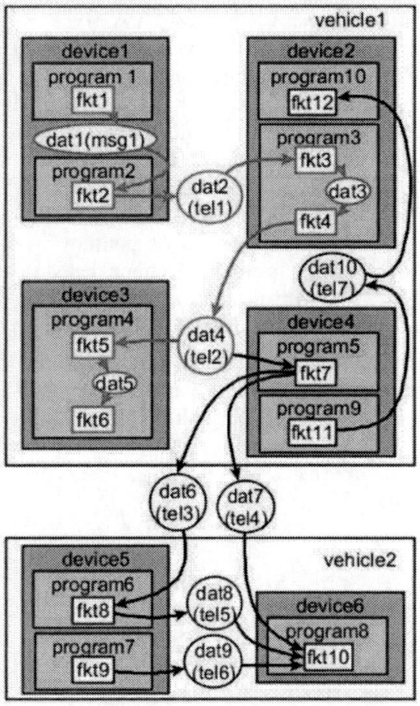

Figure 3. Petri net example of a distributed control system [10]

B. Information-Centric Networking

ICN reverses the traditional IP address-centric networking into a content centric one enabling the user to directly retrieve the content using a "name" without referring to the IP address of the node string the content.

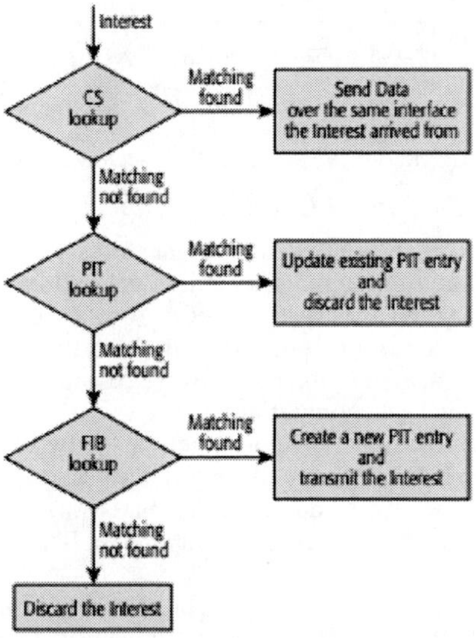

Figure 4. Named Data Networking Interest processing at an intermediate node [14]

This concept sits well with vehicular networks due to poor quality wireless links and the mobility of vehicles, data delivery is challenging. ICN-based VANET concept is presented in [14] where ICN-based VANETs can be applied in areas of application, mobility and security. For example, the VANET data will contain location, time-stamp, etc., so, if the road conditions are sought, ICN can match this better than name-to-IP-address resolution and thus the vehicle does not need to be always connected. The mobility issues in ICN are addressed by the use of named data and therefore, the anycasting and in-network caching properties of ICN allow vehicles to retrieve content from the most convenient storage point. Content-based security is supported by ICN with protection and trust implemented at the packet level rather than at the communication channel.

In Figure 4, a node follows the algorithm, it first looks in its content store (CS) to find a content copy, if a match is not found it looks in the pending interest table (PIT) and eventually in the forwarding information base (FIB). In Figure 4 the Interest/Data exchange refers to a vehicular environment.

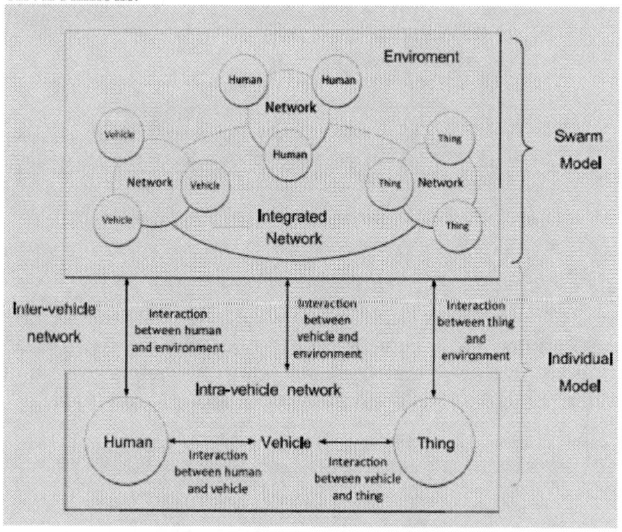

Figure 5. Swarm and Individual Network Model of IoV [9]

C. *Swarm and Individual Network Model*

The model presented in [9] integrates human, vehicle, thing and environment. The individual model focuses on one vehicle and the swarm model focuses on multi-user, multi-vehicle, multi-thing and multi-network scenarios. Through swarm intelligence, crowd sensing and sourcing and social computing, IoV can provide services/applications. Factors such as network partitions, route failures, change in channel quality and data rate and network load are addressed using swarm intelligence computing at the service providing stage. This is shown in Figure 5.

Authors in [9] also highlight that understanding the service limits is critical for sustainability i.e. network resources under diverse high-dimensional data and limited bandwidth of the wireless network.

D. *Cloud, Connection and Clients*

Three major network elements of IoV are identified in [30] as cloud, connection and client as shown in Figure 6. The 'cloud' infrastructure provides a platform for a range of wireless access technologies. With the magnitude of traffic related information likely to drastically increase, it is ideal to handle the information using cloud computing framework. 'Connection', on the other hand, utilises Third Party Network Inter Operator (TPNIO) to reduce direct Service Level Agreement (SLA) between the operators of the networks, enabling seamless roaming without compromising the quality and security of network operators. The 'client' element with the help of Wireless Access Technology (WAT) are broadly prioritized and split applications into safety and management oriented and business oriented.

Figure 6. The three network elements of IoV [30]

IV. IoV Network Models - Research Challenges & Solutions

In Section III, we have presented some of the network models from literature. These models were chosen as they show the new paradigm of ICN and the concept of how the proposed network model can communicate both with 'people' and 'things'. Petri-net was chosen as it gives the flexibility for distributed control. The challenge for any network model in IoV is to be able to exchange information from V2V and V2x, where x can be a roadside sensor, another device or a person. In addition, there may be incompatibility among devices, different qualities and response time for Internet connections and limited access to data processing and storage. There will be additional complexity where some vehicles will be connected while others not.

Future and emerging vehicle applications will consume a huge amount of sensor data in a collaborative manner. Content centric [31] and information-centric networks will play a key role. Vehicles move fast, therefore, in a content-centric networking style, vehicle position, speed and direction from the rest of the vehicles are continuously sent. Whereas, ICN focusses on what instead of where to fulfil primary demands from both content publishers and consumers. Vehicular-cloud and ICN will contribute to the 'cloud' to produce advance vehicular services, resource sharing and storing. The proposed architecture for ICN – Named Data Networking (NDN) [32] has been extended to vehicular networks where content is found and not hosts or IP addresses.

ICN seems to be the most favorable network model with distributed control. From literature, we have shown petri-nets [10] used in a number of IoV scenarios. Researchers are leading towards layered architectures [30][33] where network is one of the layers in the IoV architecture. The revealing of location information has huge concerns in vehicle privacy. In addition, location verification of neighbouring vehicle is also challenging due to the absence of trusted authority in vehicular communication. To capture vehicles in line of sight and away from sight presents yet another challenge due to the impact of moving and static obstacles in the network model.

Automobiles are undergoing a revolution by changing the way we think of the cars. Autonomous driving is trialled in developed countries, while fully autonomous driving may be a few years away, some cars will be connected in the very near future, while others not. This will bring challenges in IoV and solutions will have to account for the not connected car. A high volume of data will be exchanged in V2x and data will split between information-rich and safety-critical. Current forms of IP address centric model and control will be challenged due to the dynamic environment of the vehicular data. Therefore, information-centric networking based on distributed control and petri-nets may be the way forward. However, in the short term inter-working between existing networking technologies and information-centric network will be needed. QoS is not guaranteed currently in ICN, and to enable that, software defined networks under the umbrella of network functions virtualization and vehicular cloud networking will be the key enabler.

The integration of automotive and information technology will be promoted as a result of IoV. The biggest challenge in IoV implementation is the lack of coordination and communication. The push for IoV will generate massive data sets. Their analysis will help in the management of traffic systems and towards an intelligent transportation system. The main interactions are between the vehicle and its environment. Hence separate models can be presented in a layered architecture. Some of the challenges identified are:

- Maintaining an accurate line of sight
- Accounting for vehicles/x that are outside the line of sight
- Position/velocity of the vehicle in order to model the dynamic platoon of vehicles
- Vehicles that are not connected
- Security considerations and protection from theft
- Integration of different wireless protocols e.g. DSRC, IEEE 802.11abgn Wi-Fi, 4G/5G cellular networks, VLC
- Device-to-Device (D2D) communication (defined as direct communication between devices in range proximity without the involvement of a network infrastructure) [34] based on LTE
- Safety vs comfort applications
- Integration with cloud architecture
- Big data analysis in IoV
- QoS guarantee – investigate into SDN techniques based on the combined information from multiple sources rather than individual

V. CONCLUSIONS

This paper presents an overview of the network models for connected vehicles under the umbrella of IoV and applies the concept of NoT to IoV. IoV is emerging and will be integrated with information technology. The 'big data' generated as a result of connected vehicles will be useful in shaping the management of vehicles thus improving road safety and traffic congestion. The QoS will be split between safety critical and lower priority applications. For example, comfort subsystems within a vehicle are not safety critical. However, suspension and braking system, traction control, etc., will be prioritized and require QoS. Research in understanding existing network models is a key starting point and will enable us in establishing the modelling direction.

IoV is a building block towards the roadmap to autonomous cars. IoV will revolutionize cars like mobile phones did ten years ago. CAN communication model using petri-nets will be further examined for IoV application given its low bandwidth with high reliability benefits.

We will build on this review and the future work will focus on proposing network models for connected vehicles for IoV.

ACKNOWLEDGMENT

The work reported here is supported in part by the Santander Seed-Corn Research Grant.

REFERENCES

[1] Land transportation system report by Cisco http://www.cisco.com/c/dam/global/pt_pt/assets/cisco-connect-2013/pdf/16_outubro_helder_antunes.pdf [retrieved May 2017].

[2] J. W.ang, C. Li, H. Li and Y. Wang, "Key technologies and development status of Internet of Vehicles", 9th Int. Conf. on Measuring Technology and Mechatronics Automation, 14-15 Jan. 2017.

[3] W. Zhang and X. Xi, "The innovation and development of Internet of Vehicles", Network Technology and Application, China Comunications, pp 122-127, 13 June 2016.

[4] N. Lu, N. Cheng, N. Zhang, X. Shen and J. Mark, "Connected vehicles: solutions and challenges", IEEE Internet of Things, Journal, vol. 1, no. 4, pp 289-299, Auguat 2014.

[5] A. Dua, N. Kumar and S. Bawa, "A systematic review on routing protocols for Vehicular Ad Hoc Networks", Vehicular Communications, vol. 1, no. 1, pp. 33-52, 2014.

[6] G. Karagiannis, O. Altintas, E. Ekici, G. Heijemk, B. Jarunpan, K. Lin and T. Weil, "Vehicular networking: A survey and tutorial on requirements, architectures, challenges, standards and solution", IEEE Communications Surveys and Tutorials, vol. 13, no. 4, fourth quarter, 2011.

[7] S. K. Datta, R. Pedro, F. D. Costa, J. Harri and C. Bonnet, "Integrating connected vehicles in Internet of Things ecosystems: Challenges and solutions", IEEE 17th International symposium on a A world of woreless, mobile and multimedia networks (WoWMoM), 21-24 June 2016.

[8] M. Gerla, E-K, Lee, G. Pau and U. Lee, "Internet of Vehicles: From intelligent grid to autonomous cars and vehicular clouds", IEEE World Forum in Internet of Things (WF-IoT), 2014

[9] Y. Fangchun, W. Shangguang, l. Jinglin, L. Zhihan and S. Qibo, "An overview of Internet of Vehicles, Vehicular Networking", China Communication, vol. 11, issue 10, pp. 1-15, Oct 2014.

[10] Y. Kim, Y. Song and J. Lee, "Vehicular authentication security mechanism modelling using petri nets", Indian Journal of Science and Technology, vol. 8(S7), pp. 443-447, April 2015.

[11] P. Diekhake, T. Kurczveil and U. Becker, "Modelling and simulation of distributed time critical communication and control systems in vehicles or vehicle networks", pp. 303-308, IFAC 2016.

[12] M. D. S. Soares and J. Vrancken, "A modular petri net to modelling and scenario analysis of a network of road traffic signals", Control Engineering Practice, vol. 20, issue 11, pp. 1183-1194, Nov. 2012.

[13] Y. Li, M. Zhao and W. Wang, "Intermittently connected vehicle-to-vehicle networks: Detection and analysis", IEEE Globecom, 5-9 Dec. 2011.

[14] M. Amadeo, C. Campolo and A. Molinaro, "Information-centric networking for connected vehicles: A survey and future perspectives", Automotive Networking and Applications, IEEE Communications magazine, vol. 54, issue 2, pp. 98-104, Feb. 2016.

[15] J. Voas,."Network of 'Things'", NIST Special publication, 800-183, 2016. http://nvlpubs.nist.gov/nistpubs/SpecialPublications/NIST.SP.800-183.pdf [retrieved May 2017].

[16] S. Tuohy, M. Glavin, C. Hughes, E. Jones, M. Trivedi and L. Kilmartin, "Intra-vehicle: A review", IEEE Transactions on Intelligent Transportation Systems, vol. 16, issue 2, pp. 534-545, April 2015.

[17] G.Charalampopoulos, T. Dagiuklas and T. Chrysikos, "V2I Applications in Highways: How RSU dimensioning can improve service delivery", 23rd International Conference on Telecommunications (ICT 2016), May 16-18, 2016, Thessaloniki, Greece.

[18] Society of Automotive Engineers, "Dedicated Short Range Communications (DSRC) Message Set Dictionary", J2735 Ground Vehicle Standard, J2735, 2016.

[19] S-H. Yu, O. Shih, and H-M. Tsai, "Smart Automotive Lighting for Vehicle Safety", IEEE Communications Magazine, vol. 51, issue 12, pp 50-59, Dec. 2013.

[20] M. Wellens, B. Westphal, and P. Mahonen, "Performance evaluation of IEEE 802.11-based WLANs in vehicular scenarios", in IEEE 65th Vehicular Technology Conference, April 2007, pp. 1167–1171.

[21] D. Carona, A. Serrador, P. Mar, R. Abreu, N. Ferreira, T. Meireles, J. Matos, and J. Lopes, "A 802.11p prototype implementation", IEEE Intelligent Vehicles Symposium (IV), 2010, La Jolla, CA, USA, June 2010.

[22] L. Cheng, B. E. Henty, R. Cooper, and D. D. Stancil, "A measurement study of time-scaled 802.11a waveforms over the mobile-to-mobile vehicular channel at 5.9 GHz", IEEE Communications Magazine, vol. 46, pp. 84–91, May 2008.

[23] A. Paier, R. Tresch, A. Alonso, D. Smely, P. Meckel, Y. Zhou, and N. Czink, "Average downstream performance of measured IEEE 802.11p infrastructure-to-vehicle links", EEE International Conference on Communications (ICC) Workshops, pp.1 -5, 2010.

[24] D. Jiang and L. Delgrossi, "IEEE 802.11p: Towards an international standard for wireless access in vehicular environments," in IEEE 67th Vehicular Technology Conference, May 2008, pp. 2036–2040.

[25] S. Grafling, P. Mahonen, and J. Riihijarvi, "Performance Evaluation of IEEE 1609 WAVE and IEEE 802.11p for Vehicular Communications", International Conference on Ubiquitous and Future Networks, pp. 344-348, 2010.

[26] K. Y. Ho, P. C. Kang, C. H. Hsu, and C. H. Lin, "Implementation of WAVE/DSRC devices for vehicular communications", in Computer Communication Control and Automation (3CA), 2010 International Symposium on, vol. 2, 5-7 May 2010.

[27] IEEE webinar on connected vehicles Part I and II, Dec. 2015.

[28] T. Murata, "Petri nets:Properties, analysis and applications", Proceedings of the IEEE, vol. 77, no. 4, pp541-580, 1989.

[29] CAN Specification, Robert Bosch GmbH, Stuttgart, Germany, 1991.

[30] O. Kaiwartya, A. H. abdullah, Y. Cao, A. Altameem, M. Prasad, C-T. Lin and X. Liu, "Inter of Vehicles: Motivation, layered architecture, network model, challenges and future aspects", IEEE Access, vol. 4, pp. 5356-5373, 28 Sept. 2016.

[31] V. Jacobson, D. K. Smetters, J. D. Thornton, M. F. Plass, N. H. Briggs, and R. L. Braynard, "Networking Named Content," in ACM CoNEXT, Rome, Italy, 1-4, Dec. 2009.

[32] S. H Ahmed, S.H. Bouk, M. A. Yaqub, D. Kim, H. Song, and J. Lloret, "CODIE: Controlled data and interest evaluation in vehicular named data networks," IEEE Trans. Veh. Technol., vol. 65, no. 6, pp. 3954_3963, Jun. 2016.

[33] J. Contreras-Castillo, S. Zeadally and J. A. G. Ibanez, "A seven-layered model architecture for Internet of Vehicles", Journal of Information and Communication, vol. 1, issue 1, pp. 4-22, March 2017.

[34] T. T. Gunes, S. T. Kwaong U and H. Afifi, "Hybrid model for LT-assisted D2D communications", Ad-hoc, Mobilr and Wireless Networks conference (AdhocNow), vol. 8487, pp. 100-113, 2014.

Enhancing QoS in Vehicle-to-Infrastructure Approaches through Adaptive Cooperative Communications

Radwa Ahmed Osman[*#], Xiao-Hong Peng[*]and M. A. Omar[#]
[*]School of Engineering & Applied Science, Aston University
Birmingham, UK
[#]College of Engineering & Technology, Arab Academy for Science, Technology & Maritime Transport
Alexandria, Egypt
email:{osmanmra, x-h.peng}@aston.ac.uk, mabd_elhamid@aast.edu

***Abstract*— In vehicle-to-infrastructure (V2I) communications, the challenge regarding how to achieve the intended performance at the minimal resource cost has yet been well addressed in the recent development of vehicular technologies. In this paper, we investigate some data transmission schemes, such as cooperative communications, for improving quality of service (QoS) in vehicular networks. We propose a method that facilitates V2I through vehicle-to-vehicle (V2V) communications and, for this approach, we derive the closed-form expressions of the outage probability, throughput, energy efficiency and packet loss rate for four different transmission schemes investigated. The QoS performances can be optimized by finding appropriate transmission schemes with a certain number of relays within a given transmission distance. The proposed approach is also aimed to achieve the best performance trade-off between system reliability and efficiency under various environmental conditions.**

***Keywords*— QoS; V2I; V2V; cooperative communications.**

I. Introduction

In Vehicle-to-Infrastructure (V2I) communication networks, mobile users are able to access Internet services such as traffic condition broadcast, video streaming, digital map downloading, and information of road hazard and accident alarm, via fixed roadside units. The most recent research in this area has been focused on the vehicular ad-hoc networks (VANET) [1] [2], including its connection to the Fourth-Generation or Long-Term Evolution (LTE and LTE-Advance) cellular networks and the provision of good solutions to V2I in order to ensure low latency and high reliability communication [3].

IEEE 802.11.p is one of the commonly used standards for V2I to support vehicular communications in highly mobile, often densely populated, and frequently non-line-of-sight environments [4] [5]. In addition, the IEEE 802.15.4 standard, comprising a simple physical (PHY) layer and an energy efficient medium access control layer, is also designed to support both real-time and contention-based services and has been considered as a promising candidate for Internet of vehicles (IoV) and vehicular sensor networks [6]. To tackle the problem of high packet loss rate the cooperative communications techniques can be applied to enhance transmission reliability by creating diversity [7]. In this case, mobile nodes (vehicles) can help each other through relaying other node's data and sharing their limited resources, to improve loss performance and increase transmission coverage. However, the performance enhancement by using relays nodes is constrained by the power (energy) budget imposed and high mobility in the vehicular network [8]. This issue can potentially impede the delivery of quality of service (QoS) in the V2I approach.

In this work, we investigate both cooperative and non-cooperative transmission schemes, and intend to reveal how these schemes perform in the context of a vehicular network, in terms of energy consumption, throughput and packet loss rate under different conditions, such as transmission distance, relaying method and channel condition (path loss exponent). These findings are used to identify proper transmission schemes that can optimize the system performance for the whole network in a changing environment. The proposed approach is unique in the sense that it provides an efficient way to find the best transmission method for transmission between any V2I links. We propose a method that facilitates V2I, which is assisted by vehicle-to-vehicle (V2V) communications when needed, and evaluate the performance of this approach based on the models we derive.

The remainder of this paper is organized as follows. Section II discusses the relevance of this research with other work. The system models for both cooperative and non-cooperative transmission schemes for V2I communications are presented in Section III. Simulation results produced by Matlab and NS-2 and discussions are presented in Section IV. Finally, the paper is concluded in Section V.

II. Related Work

Cooperative communications have been studied extensively for VANETs and two of the most common protocols of this technology are Amplify-and-Forward (AF) and Decode-and-Forward (DF) [9]. Cooperative or polarization diversity is created in these protocols through exploiting the broadcast nature of wireless channels and using relays to improve link reliability and throughput in a vehicular network [10]. In addition, the use of graph theory to formulate the problem of cooperative communications scheduling in vehicular networks is proposed in [11], in order to improve the throughput and spectral efficiency of vehicular networks.

Enhancing system efficiency is a key issue in applying cooperative communications in V2I approaches, depending on the connectivity probabilities in V2I and V2V communication scenarios in one-way and two-way platoon based VANETs [12]. Smart Antenna technology can also contribute to the increment of the service coverage and system throughput of V2I [13]. The capacity of V2I communications can be maximized by an iterative resource-allocation method [14] and the efficiency of V2I communications can be improved by applying a scheme called Distributed Sorting Mechanism (DSM) [15]. To improve power efficiency in vehicle-to-roadside infrastructure (V2I) communication networks, [16] proposed a joint power and sub-carrier assignment policy under delay-aware QoS requirements. In addition, the strong dependence on the environment due to multipath propagation is also presented for a distributed energy efficient routing method [17].

Most of the works have demonstrated the possibility of improving the system performance of vehicular networks by using different methods. However, there is a lack of information regarding how to choose a specific transmission scheme under different conditions in terms of the number of relaying branches and the number of relays for a given distance between source and destination nodes, in order to find a solution for ensuring the best QoS.

In this paper, our focus will be the identification of the conditions for establishing appropriate transmission strategies among different commonly used transmission schemes, including both cooperative and non-cooperative schemes for V2I communications. Our approach is based on the development of analytical models for these transmission schemes and the assessment of their performances in reliability, energy efficiency and throughput. It also reveals the trade-offs between cooperative and non-cooperative transmission schemes and shows how to utilize this property to achieve the optimized performance through adaptive cooperative communications.

III. SYSTEM MODEL

In this section, the analytical models of the required transmitting power, outage probability, energy efficiency, throughput and packet loss rate in the context of a V2I network are established for both cooperative and non-cooperative transmission schemes. Based on these models, an adaptive transmission strategy can be developed to optimize the system performance.

Given a V2I network with N vehicles, for any vehicle-to-infrastructure pair (V, I), where $V \epsilon \{1, \ldots, N\}$, the goal of optimizing the transmission QoS is achieved by minimizing the total energy consumed per bit (or energy efficiency) with a given outage probability target, maximizing the end-to-end throughput, and minimizing the packet loss rate based on the transmission distance between V2I pairs, i.e.

$$Min \sum E_{bi} \quad s.t. \{p_{outVI}\} \text{ and}$$
$$Max \sum S_{thi} \quad s.t. \{d_{VI}\} \qquad (1)$$

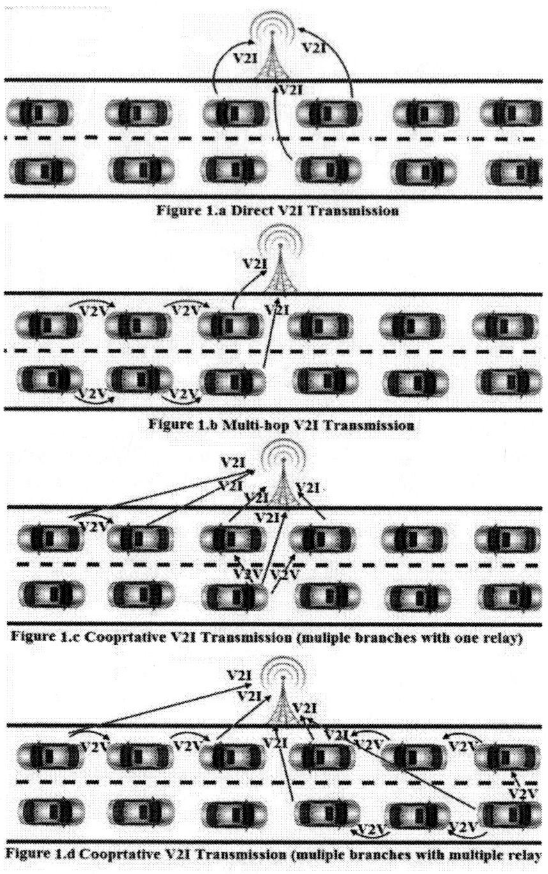

Figure 1.a Direct V2I Transmission

Figure 1.b Multi-hop V2I Transmission

Figure 1.c Cooprtative V2I Transmission (muliple branches with one relay)

Figure 1.d Cooprtative V2I Transmission (muliple branches with multiple relay

Figure1. Different V2I transmission schemes.

where E_{bi} and S_{thi} are the energy consumed per bit and throughput, respectively, of the i-th path between a vehicle (V) and infrastructure (I), p_{outVI} and d_{VI} are the fixed outage probability target and the total transmission distance between V and I.

Four transmission schemes in the context of V2I are identified in Figure1, including single-hop direct V2I (1a), multi-hop V2I via V2V (1b), cooperative V2I with a single relay (1c), and cooperative V2I with multiple relays (1d). In this work, we intend to examine the performances of different transmission schemes in terms of energy efficiency, throughput and packet loss rate, and to optimize them under different environmental conditions.

We consider a V2I network in which the transmission links are subject to narrowband Rayleigh fading with additive white Gaussian noise (AWGN) and propagation path-loss. The channel fades for different links are assumed to be statistically mutually independent. For medium access, vehicle nodes are assumed to transmit over orthogonal channels through using the service channels specified in IEEE801.11p [2], thus no mutual interference is considered in this system model. These channels can be reused by other vehicle away from a certain distance.

A. *Non-Cooperative Transmission Scheme*

Consider the transmission scheme for a direct link (V, I) as shown Figure 1a where no relaying paths are involved. We use P_{SDir} to denote the source transmission power for this case. For the direct transmission in the V-I link, the received symbol r_{VI} and the spectral efficiency R can be modeled as:

$$r_{VI} = \sqrt{P_{SDir}\, d_{VI}^{\ \alpha}}\, h_{VI} s + N_o \tag{2}$$

$$R = \frac{1}{2}\log_2\left(1 + SINR_{VI}\right) \tag{3}$$

where d_{VI} is the distance and h_{VI} is the channel coefficient of the V-I link, α is the path loss exponent, s is the transmitted symbol with unit power and N_o is the Gaussian noise.

The log-normal environment shadowing path loss model at a distance d_{ij} rom node i and node j is given by [18]:

$$\gamma_{ij}[dB] = PL(d_o) + 10\alpha\log_{10}\frac{d_{ij}}{d_o} + X_\sigma \tag{4}$$

where X_σ is a zero-mean Gaussian distributed random variable with standard deviation σ and with some time correlation. This variable is zero if no shadowing effect exists. The $PL(d_o)$ is the path loss at a reference distance d_o in dB. The Signal-to-Noise Ratio (SNR) of the V-I link is:

$$SINR_{VI} = \frac{P_{SDir}|h_{VI}|^2\gamma_{VI}}{N} \tag{5}$$

where $N = N_o B$ is the noise power spectral density, and B is the system bandwidth in Hertz.

An outage occurs when the SNR at the receiver falls below a threshold β which allows error free decoding. This threshold is defined as $\beta = 2^{2R_s} - 1$, where R_s is the required system spectral efficiency. The outage probability of the single-hop transmission is given by [19]:

$$p_{outVI} = p(SINR_{VI} \le \beta) = 1 - e^{\left(-\frac{(2^{2R_s}-1)N}{P_{SDir}|h_{VI}|^2\gamma_{VI}}\right)} \tag{6}$$

Energy consumption is largely proportional to the requirement of maintaining a certain level of transmission reliability or the successful transmission rate. In order to maintain a required level of reliability, denoted by U, which is related to the reliability of a transmission link, the minimum outage probability is defined as:

$$p_{out} \le 1 - U \tag{7}$$

Combining (6) and (7) and taking the nature logarithm on the both sides of expression, we have:

$$\frac{(2^{2R_s}-1)N}{P_{SDir}|h_{VI}|^2\gamma_{VI}} \le \ln(U^{-1}) \tag{8}$$

The main objective for the performance optimization of a V2I network is to minimize the total energy consumption under different environmental conditions. Thus, the transmit power required to satisfy the reliability requirement or be constrained by the outage probability for the direct transmission must be:

$$P_{SDir} \ge (2^{2R_s} - 1)\frac{N}{|h_{VI}|^2\gamma_{VI}}\left(\ln(U^{-1})\right)^{-1} \tag{9}$$

Therefore, the total consumed energy per bit (J/bit) for the direct transmission mode can be expressed as:

$$E_{bDir} = \frac{P_{AM,Dir} + P_C}{R_b}, \quad where\, P_C = P_{Tx} + P_{Rx} \tag{10}$$

$$P_{AM,Dir} = \frac{\xi}{\eta}P_{SDir} \tag{11}$$

where $P_{AM,Dir}$ is the power amplifier consumption for direct transmission which depends on the drain efficiency of the amplifier η, the average peak-to-peak ratio ξ, and the transmit power P_{SDir}, $R_b = R_s B$ is the data rate in bits/s, B is the system bandwidth, Pc is the power consumed by the internal circuitry for transmitting (P_{Tx}) and receiving (P_{Rx}).

The throughput S_{th} and packet loss rate PL can simply be defined, i.e.

$$S_{th} = \frac{Total\,Received\ Payload}{Total\,Transmitted\ Time} \tag{12}$$

$$PL = \frac{Total\,Sent\ Packets \quad Total\,Received\ Packets}{Total\,Sent\ Packets} \tag{13}$$

The multi-hop non-cooperative transmission scheme with n ($n \ge 1$) relays is shown in Figure 1b. Each relay is able to detect if the packet was received correctly or not and will forward the information to the destination only in the case of the packet being correctly received. Otherwise, the packet is considered lost.

Given the outage probability of individual hops, i.e., p_{outVR_1} (from a vehicle to relay 1), $p_{outR_1R_2}$ (from relay 1 to relay 2), ..., p_{outR_nI} (from relay n to infrastructure), the outage probability of the multi-hop link, p_{outMH}, is given by:

$$p_{outMH} = 1 - (1 - p_{outVR1})(1 - p_{outR1R2})\ldots(1 - p_{outRnI}) \tag{14}$$

With the same mathematical treatment as in (6), p_{outMH} becomes:

$$p_{outMH} = 1 - e^{\ (2^{2R_s}-1)N}\, y \tag{15}$$

$$where\ \ y = \left(\frac{1}{P_{Vr1}|h_{Vr1}|^2\gamma_{Vr1}} + \sum_{i\ 2}^{n}\frac{1}{P_{r1ri}|h_{r1ri}|^2\gamma_{r1ri}} + \frac{1}{P_{rnI}|h_{rnI}|^2\gamma_{rnI}}\right)$$

We set the transmit power to be proportional to the distance between two communicating nodes. For broadcast transmission, e.g., when the source transmits, the longest distance, i.e. the distance between the source and the destination d_{sd}, is considered. The power minimization problem is specified in a similar way to (7) and the total consumed energy per bit for the multi-hop direct transmission is expressed:

$$E_{bMH} = (p_{outMH})\frac{P_{AM,MH} + P_C}{R_b} + (1 - p_{outMH})\frac{(n * X + 1)P_{AM,MH} + (n+1)P_C}{R_b} \quad (16)$$

where $P_{AM,MH}$ is the power amplifier consumption for multi-hop transmission.

B. Cooperative transmission Scheme

In cooperative transmission, the sender V broadcasts its symbol in to all potential receivers including the destination I and relays in the current time slot. Two types of cooperative transmission schemes are considered here: 1) using multiple cooperative relaying branches with one relay in each branch (Figure 1c), and 2) multiple relaying branches with multiple relays in each branch (Figure 1d). The selective decode-and-forward (SDF) relaying protocol is used in these two schemes and relays perform cooperation when the information from the source is correctly received by them. Based on the derivations methods used in Subsection III. A, the following close-form expressions can be readily obtained.

1): The outage probability of cooperative transmission with multiple (K) relaying branches:

$$p_{outMB} \approx (2^{2R_s} - 1)^{K+1} N^{K+1} z \quad (17)$$

where $z = \left(\frac{1}{P_{VI}|h_{VI}|^2 \gamma_{VI}} \left(\frac{1}{P_{VI}|h_{Vr}|^2 \gamma_{Vr}} + \frac{1}{P_{rI}|h_{rI}|^2 \gamma_{rI}} \right)^K \right)$

2): The lower bound of power for the SDF scheme:

$$P_{SMB} \geq (2^{2R_s} - 1) N(y) \frac{1}{K+1} (\ln(U^{-1}))^{\frac{1}{K+1}} \quad (18)$$

3): The total consumed energy per bit:

$$E_{bMB} = (p_{outVr})\frac{P_{AM,MB} + P_{Tx} + (K+1)P_{Rx}}{R_b} + (1 - p_{outVr})\frac{(K * X + 1)P_{AM,MB} + (K+1)P_{Tx} + (K+2)P_{Rx}}{R_b} \quad (19)$$

4): The total consumed power:

$$P_{totMB} = (p_{outVr})(P_{AM,MB} + P_{Tx} + (K+1)P_{Rx}) + (1 - p_{outVr})((K * X + 1)P_{AM,MB} + (K+1)P_{Tx} + (K+2)P_{Rx}) \quad (20)$$

The transmit power at relays can be reduced and consequently the energy efficiency will be improved by implementing the cooperative communications schemes, which are particularly suitable for long-range transmission - the related results will be shown in Section IV.

IV. NUMERICAL RESULTS AND DISCUSSION

In this section, we examine the performances of different transmission schemes through Matlab and NS-2 simulations in terms of energy efficiency (energy consumption per bit), throughput and packet loss rate. We then reveal the conditions for selecting the optimal transmission schemes through comparisons between them. The network settings used for simulation is listed in TABLE 1. Assume the spectral efficiency Rs in this scenario to be 2 bit/sec/Hz, and the required system reliability level to be 0.999. To generate mobility, mobility-files are created in ns-2 simulation. In addition, we assume that all the vehicles are running at the same speed and keeping the same distance with each other.

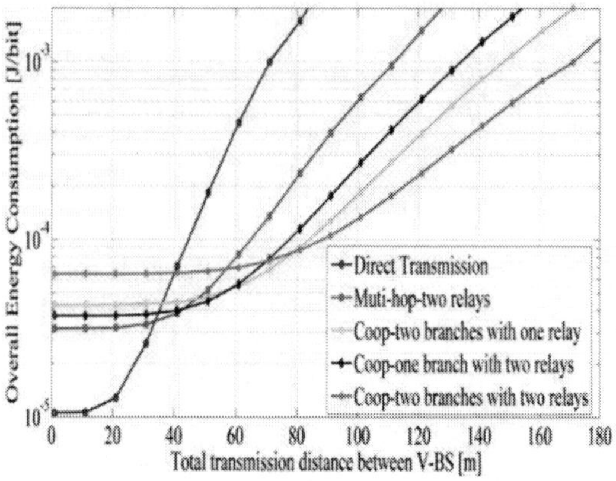

Figure 2. Total energy consumed vs total transmitted distance.

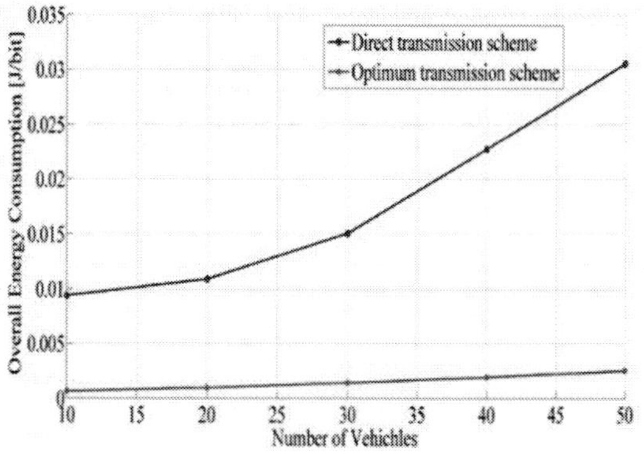

Figure 3. Overall energy consumption vs number of vehicles

In Figure 2 the energy performances of both cooperative and non-cooperative schemes are illustrated and compared. As we can see, the non-cooperative direct transmission has the lowest energy cost than all others transmission schemes for short-range (d_{VI}<33m); the non-cooperative transmission using multi-hop relays outperforms the direct transmission for the range 33m<d_{VI}<43m and, in particular, transmission using two intermediate relays (n=2) nodes has the lowest energy consumption for this range.

The cooperative transmission outperforms the non-cooperative transmission schemes for the range 43m<d_{VI}<58m, and the transmission using one branch with two relays (K=1, n=2) has the lowest energy consumption for this range. As distance continuously increases, the lowest energy consumption is achieved by transmission using two branches with one relay (K=2, n=1) for 58<d_{VI}< 80m, and by transmission using two branches with two relays (K=2, n=2) for d_{VI}>80m.

TABLE 1. SIMULATION PARAMETER

Parameters	Value
N_0	−174 dBm
B	10 kHz
R_s	2 bit/sec/Hz [20].
P_{TX}	97.9 mW [20]
P_{RX}	112.2 mW [20]
η	0.35
ξ	0.5
Packet Size	512 bytes
f_c	5.9 GHz
α	3
Simulation time	1000 Sec
Nodes	10/20/30/40/50
Velocity	5 km/h, 20 km/h, 60 km/h
Traffic Agent	TCP
Mac Protocol	IEEE 802.11p
Queue	PriQueue with size of 50 Packets
Propagation model	Log-normal shadowing Model (LOS)
Antenna	Omni-directional with height of 1m
Routing Protocol	AODV
Number of Seed	3

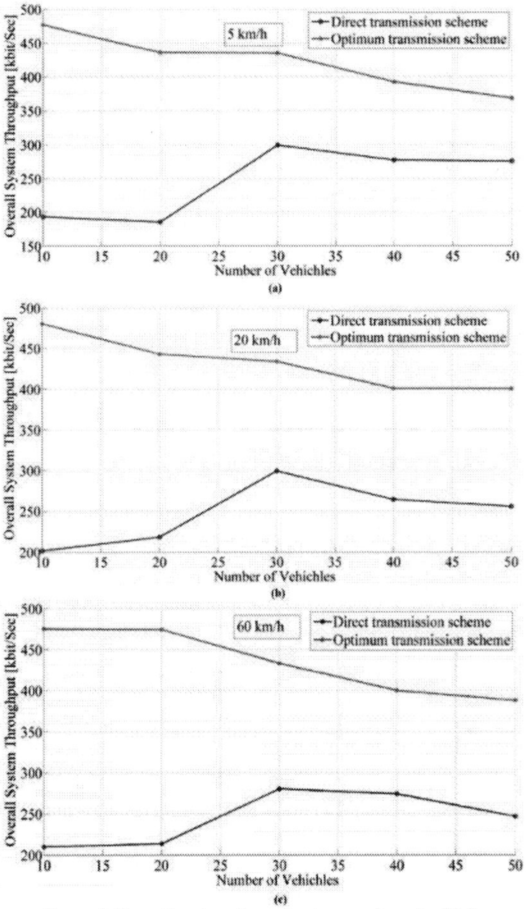

Figure 4. Overall system throughput vs number of vehicles.

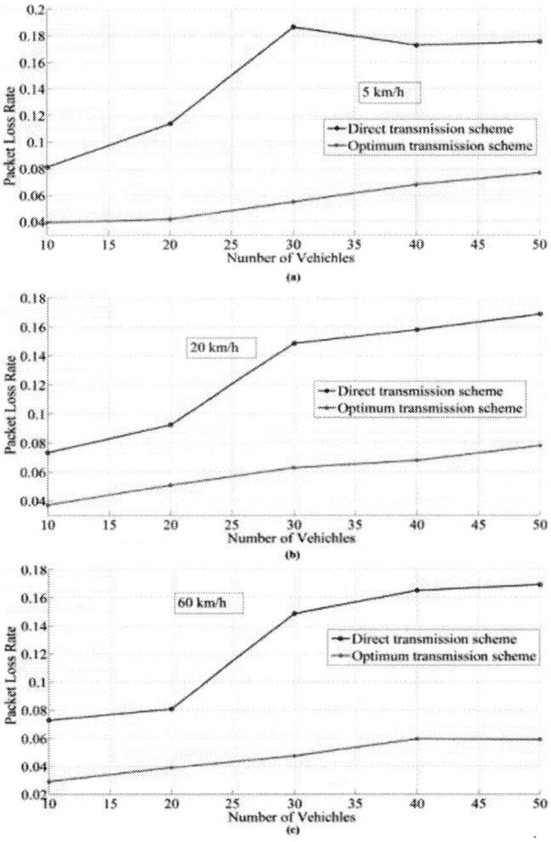

Figure 5. Packet loss rate vs number of vehicles

As shown in Figure 3, the non-cooperative direct transmission has much higher energy consumption than the optimum transmission scheme which is chosen based on the transmission distance between vehicles and infrastructure.

The overall system throughput is shown in Figure 4 for three different vehicle velocities. The optimum transmission schemes clearly outperform the direct transmission schemes in all cases due to the impact of diversity created by cooperative transmission. It is also noticed that the throughput of the optimum transmission scheme decreases when the number of transmitting vehicles increases. This is mainly due to congestion in medium access and increased operation overhead at the nodes that are the source, as well as the relay at the same time.

Figure 5 depicts the overall system packet loss rate for direct transmission and optimum transmission schemes versus the number of transmitting vehicles for different vehicle velocities. As it is shown, the packet loss rate increases when the number of transmitted vehicles increases for all the transmission schemes, which is caused by network congestion and correlated with the corresponding performance in throughput as shown in Figure 4. It is worth mentioning that the optimum transmission schemes have much lower packet loss rates than the direct transmission schemes as when relays are used the transmission distances between adjacent nodes are reduced and, at the same time, the transmission reliability is improved due to the diversity generated in cooperative communications.

Due to the scenario settings in our work where most vehicles have a fairly large distance between them and the roadside base station, no major difference in performance is observed when increasing the velocity of vehicles, as shown in Figures 4 and 5. In contrast, as discussed above, the performance such as throughput is correlated with the number of vehicles which are connected to the same base station,

There are a number of factors affecting energy consumption, throughput and packet loss rate in V2I networks. Cooperative transmissions utilize additional paths and intermediate nodes which may cost more energy, but the diversity it creates can save energy by reducing the probability of link failure and consequently reducing the number of retransmissions. Diversity increases with the number of relay branches used but this increase could be marginal when the number of branches is large as it is difficult to ensure that all the branches are uncorrelated.

Clearly, to achieve the best energy performance as discussed in this paper, proper transmission schemes should be selected for the given transmission conditions such as the overall distance, d_{Sd}, and channel quality in terms of α. The findings of this work can assist deciding when and how the cooperative or non-cooperative transmission scheme should be employed. Based on our investigation, an energy-efficient transmission strategy can be formed in a V2I network by adaptively choosing proper transmission schemes under different network and transmission conditions. This involves determining the number of relaying branches and the number of relays if the cooperative scheme is to be used. By doing so, energy saving could be significant even with the direct transmission scheme in certain conditions, as shown from our results.

V. Conclusion

We have investigated different transmission schemes in terms of their energy and throughput performances for V2I communications. We have shown that both cooperative and non-cooperative transmission schemes can exhibit the best performance under certain environmental conditions. The optimal transmission scheme can be identified given the distance between the source and destination nodes in a V2I network. The results presented in this paper can be used to form an adaptive transmission strategy that is able to select an appropriate transmission scheme in a changing environment to maintain the best QoS performance in a dynamic way, in terms of achieving the highest throughput with a fixed energy budget or the lowest energy cost for the given throughput target.

References

[1] S. Wu, C. Chen, and M. Chen, "An asymmetric and asynchronous energy conservation protocol for vehicular networks," IEEE Trans. Mobile Compu., Vol. 9, pp. 98-111, January 2010.

[2] M. J. Booysen, S. Zeadally, and G.-J. V. Rooyen, "Survey of media access control protocols for vehicular ad hoc networks," IET Commu., Vol. 5, pp. 1619-1631, March 2011.

[3] G. el mouna Zhioua, N. Tabbane, H. Labiod, and S. Tabbane, "A fuzzy multi-metric QoS-balancing gateway selection algorithm in a clustered VANET to LTE-advanced hybrid cellular network," IEEE Trans. on Veh. Technology, Vol. 64, pp. 804 - 817, February 2015.

[4] O. Goubet, G. Boudic, F. Gabry, and T. J. Oechtering, "Low-complexity scalable iterative algorithms for IEEE 802.11p receivers," IEEE Trans. on Veh. Technology, vol. 64, pp. 3944-3956, September 2015.

[5] X. Wu, S. Subramanian, R. Guha, R. G. White, J. Li, K. W. Lu, A. Bucceri, and T. Zhang, "Vehicular communications using DSRC: challenges, enhancements, and evolution," IEEE journal on Selected Area in Commun., Vol. 31, pp. 399-408, September 2013.

[6] C. Wang, C. Chou, P. Lin, and M, Guizani, "Performance evaluation of IEEE 802.15.4 non beacon-enabled mode for internet of vehicles," IEEE Trans. on Intelligent Transportation Systems, vol. 16, pp. 3150-3159, December 2015.

[7] A. Nasri, R. Schober, and I. F. Blake, "Performance and optimization of amplify-and-forward cooperative diversity systems ingeneric noise and interference," IEEE Trans. Wireless Commun.,vol. 10, pp.1132-1143, January 2011.

[8] D. Xie, W. Wei, Y. Wang, and H. Zhu, "Tradeoff between throughput and energy consumption in multirate wireless sensor networks," IEEE. Sensor Journal, Vol. 12, pp. 3667 - 3676, August 2013.

[9] H. Hakim, H. Bouijemaa, and W. Ajib, "Single relay selection schemes for broadcast networks," IEEE Trans. Wireless Commun.,vol. 12, pp.2646-2657, May 2013.

[10] S. Sohaib and D. K. C. So, "Asynchronous cooperative relaying for vehicle-to-vehicle communications," IEEE Trans. on Commun.,Vol. 61, pp.1732-1738, May 2013.

[11] K. Zheng, F. Liu, Q. Zheng, W. Xiang, and W. Wang, "A graph-based cooperative scheduling scheme for vehicular networks," IEEE Trans. Vehicular Technology vol. 62, pp. 1450-1458, May 2013.

[12] C. Shao, S. Leng, Y. Zhang, A. Vinel, and M. Jonsson, "Performance analysis of connectivity probability and connectivity-aware MAC protocol design for platoon-based VANETs," IEEE Trans. on Veh. Technology, vol. 64, pp. 5596-5609, December 2015.

[13] S. Pyun, H. Widiarti, Y. Kwon, J. Son, and D. Cho, "Group-based channel access scheme for a V2I communication system using smart antenna," IEEE Commun. Letter, vol. 15, pp.804-806, August 2011.

[14] S. Pyun, W. Lee, and D. Cho, "Resource allocation for vehicle-to-infrastructure communication using directional transmission," IEEE Trans. on Int. Trans., vol. 17, pp. 1183-1188, April 2016.

[15] T. Wu, S. Guizani, W.-T. Lee, and K. Liao, "Improving RSU service time by distributed sorting mechanism," Ad Hoc Netw., vol. 10, pp. 212-221, March 2012.

[16] A. Goldsmith, Wireless Communications, 1st edition. Cambridge University Press, 2005.

[17] A. S. Ibrahim, Z. Han, and K.J.R. Liu, "Distributed energy-efficient routing in wireless networks," IEEE Trans. Wireless Commun., vol. 7, pp. 3930 - 3941, October 2008.

[18] T. Nguyen, O. Berder, and O. Sentieys, "Energy-efficient cooperative techniques for infrastructure-to-vehicle communications," IEEE Trans. on Int. Trans. Sys., vol. 12, pp. 659-668, September 2011.

[19] H. Zhang, Y. Ma, D. Yuan, and H-H. Chen, "Quality-of-service driven power and sub-carrier allocation policy for vehicular communication networks," IEEE Journal on Selected Area in Commun., vol. 29, pp. 197-206, January 2011.

[20] L. Azpilicueta, C. Vargas-Rosales, and F. Falcone, "Deterministic propagation prediction in transportation systems," IEEE Vehicular Tech. Mag., vol. 11, pp. 29-37, September 2016.

CBL: A Clustering Scheme for VANETs

Lucas Rivoirard
Univ Lille Nord de France,
IFSTTAR, COSYS, LEOST
F-59650 Villeneuve d'Ascq, France
lucas.rivoirard@ifsttar.fr

Martine Wahl
Univ Lille Nord de France,
IFSTTAR, COSYS, LEOST
F-59650 Villeneuve d'Ascq, France
martine.wahl@ifsttar.fr

Patrick Sondi
Univ. Littoral Côte d'Opale,
LISIC - EA 4491
F-62228 Calais, France
patrick.sondi@univ-littoral.fr

Marion Berbineau
Univ Lille Nord de France,
IFSTTAR, COSYS
F-59650 Villeneuve d'Ascq, France

Dominique Gruyer
IFSTTAR, COSYS, LIVIC
F-78000 Versailles, France

***Abstract*—Routing protocols for vehicular ad hoc networks resort to clustering to optimize network performance. In existing proposals, cluster-heads are chosen based on various metrics such as the number of its direct neighbors, the quality of the links, *etc*. Other clustering techniques consider the geographic environment of the roads, and they choose one cluster-head for each space subdivision. The clustering scheme proposed in this work combines the information on road configuration, vehicle mobility and link quality in order to build a structure similar to vehicular network infrastructure, while relying only on the vehicles. The evaluations show that this scheme allows creating and maintaining during a significant time a small number of stable connected groups, in most cases, just one in each traffic direction. This clustering scheme can be integrated into any reactive, proactive, or geographic ad hoc routing protocol in order to optimize the flooding and simplify route maintenance. And it allows the routing protocol to operate without any global location information service.**

Keywords–Clustering; Routing protocols; Cooperative vehicles; V2V; VANET; Performance evaluation.

I. INTRODUCTION

With the wide deployment of 3G/4G infrastructures, new mobile services are proposed to drivers through smartphones or built-in car devices. The concept of the connected vehicle and that of the autonomous vehicle are now effective at least in real-world testing. The development of both the IEEE 802.11p and the upcoming 5G includes machine-to-machine direct communications, which also refers to vehicle-to-vehicle (V2V) or vehicle-to-infrastructure (V2I) communications. In this context, enhancing vehicular ad hoc network (VANET) routing protocols is mandatory to ensure that this paradigm will play its role in future intelligent transportation systems. The routing protocol can lead to either a flat topology, without hierarchy [1], or a clustered topology, with a node hierarchy. Clustering is one important technique that can be used by ad hoc routing protocols to optimize network management. A cluster is a virtual division of the network into groups. Resorting to clusters optimizes the range of packet flooding by limiting the packet transmission to one or more clusters.

A cluster includes different types of nodes. The group leader is called "cluster-head". Its "ordinary member" nodes are connected to the cluster-head. "Gateway nodes" are members of several clusters, thus making a link between them. The clustering schemes can generate separated clusters (without gateway nodes) or not. Clustering methods are active, passive, or hybrid. In active clustering, dedicated control messages are sent for cluster management. In passive clustering, clusters are created on demand when data need to be transmitted. In hybrid clustering, information needed for cluster management is added to the packets. The clustering size is also characterized by the number of hops. For instance, in one-hop clusters, each member node is directly connected to its cluster-head.

Over two decades, many clustering schemes have been proposed in order to enhance the performance of ad hoc routing protocols according to various link or node metrics [2]–[4]. Especially in the case of VANET, the road traffic environment and the velocity due to vehicle mobility are important factors in the design of a clustering scheme. In order to evolve from the plethora of existing approaches towards standardized solutions, the European Telecommunications Standards Institute (ETSI) has recently published the Geonetworking requirements [5] that fix the design guidelines of VANET architecture.

This work presents an approach inspired by infrastructure-based vehicular networks. The proposed hybrid clustering scheme leads to an emerging structure, a virtual backbone in the VANET, similar to that obtained with Road Side Units (RSUs) deployed along the roads equipped for vehicular communications. In this way, any routing protocol that uses this clustering scheme can operate in both infrastructure-based and infrastructure-less VANET. When operating in infrastructure-less mode, the structure built by the proposed clustering scheme aims to offer a stable link service between the nodes for the applications, despites the high mobility of the VANET nodes. In addition, the CBL clustering scheme uses only the position and velocity information of the closest node neighborhood. Therefore, unlike most the geographic-based routing protocols, no global knowledge of the locations of the nodes is needed, and any global location service that would require an infrastructure is not necessary. The structure built by CBL allows the routing protocol supplying unicast and broadcast message exchange.

The paper is organized as follows. Section II presents a related work on clustering schemes for ad hoc networks, especially VANETs. Section III depicts the clustering scheme designed for VANET in this work. Section IV deals with its performance evaluation. We finally conclude.

II. RELATED WORK

Many clustering methods for mobile applications were first studied for Mobile Ad hoc NETworks (MANETs) [6]–[10]. The Lowest ID and the Highest Degree methods [6] offer good performance in MANET, but they do not consider mobility. An attempt to introduce mobility in the cluster-head election by considering the distances between the nodes is achieved through the Mobility Based Clustering method [8]. However, in highly dynamic ad hoc networks, it is necessary to combine several criteria, which leads to the proposal of the Weighted Clustering Algorithm [9] based on a weighted sum of several criteria values. Current studies introduce the notion of communication interest among devices as an extra parameter in the Weighted Clustering Algorithm, such as in [11] using the communication interest among MANET nodes, the physical proximity, and the energy availability to create coalitions. MANET nodes are distributed randomly in the space without favoring a geographical direction. But, VANET nodes are subject to mobility constraints such as road infrastructure with specific rules, strong variations in relative speeds between the vehicles, *etc*. Therefore, the studies that adapt clustering schemes to VANETs pay attention to the development of relatively time-stable clusters. Two main approaches can be found. The first one does not introduce logical relations between the clusters, and is based on a cost function [12]–[18], or a fuzzy logic function [19]. The second approach creates logical connections between the clusters through an optimal set of relay nodes [20]–[23], defined as a backbone, in order to improve the traditional forwarding scheme. However, most of these studies focus on the way the cluster-head is chosen at network level, without considering the special features of road applications in a global system approach.

With this in mind, the Dynamic Backbone Assisted protocol [20] uses either the distance or the communication rate as the metric to elect the cluster-head; [22] the distance; the BackBone Routing protocol [21] the speed, the traffic direction, and the quality of transmission; the Connected Dominating Set-Stable Virtual Backbone [23] the speed, the distance, and the direction. [23] searches for cluster-heads having low speeds to stabilize the chain structure. [20]–[22] consider the transmission range R as a relevance zone where there is a cluster-head [20], or a cluster-head between two associated cluster nodes located at borders of R to enable communication with two other clusters [21], or four cluster-heads distributed every $\frac{1}{4}R$ [22]. [23] puts no constraint in terms of number of cluster-heads. Furthermore, [21] introduces the notion of upstream and downstream message transmissions.

The scheme proposed in this paper will lead to the construction of a single backbone in each traffic direction, formed with cluster-heads that are dynamically chosen by the other nodes in order to form and stabilize a structure over time. As in [23], no condition will be put on the number of cluster-heads. Unlike [20]–[23], no thresholds other than temporal thresholds will be applied. Unlike [11], to obtain application-independent backbones and to preserve the collaboration between nodes, the idea of communication interest among vehicles or devices of vehicles is not considered. Also, the notions of upstream and downstream relays [18] will be exploited.

III. CBL CLUSTERING ALGORITHM

The Chain-Branch-Leaf (CBL) clustering scheme designed in this work builds a backbone that will allow both the communications in close neighborhood, and remote communications between distant vehicles according to the specific needs of each application. Communication between close vehicles will be necessary for on-board functions (sensor functions, geo-localization, extended perception, *etc.*) that will have to share variables periodically (speed, acceleration, positioning information, *etc.*) for their inner process. Such variables will be useful to coordinate the relative movements of vehicles in future autonomous systems. Communication over long distance will be necessary for distributed applications that need to forward messages, upstream or downstream the traffic flow, to remote vehicles. As an example, forwarding messages upstream can help prevent a risk of bottleneck, and transmitting messages downstream can allow vehicles to inform about the approach of a priority vehicle (police car, fire truck).

A. Assumptions

We assume that each vehicle is equipped with a GPS that enables self-localization, also that it can determine its speed. It also has a wireless ad hoc communication card (802.11p for an example) enabling communication with the other vehicles up to a certain range in line-of-sight.

B. Definitions

CBL is a hybrid distributed algorithm: each communication node initiates its own process. It creates a hierarchy between the nodes in order to build 1-hop clusters so that each node of a cluster can directly communicate to the cluster-head without going through another intermediary node. CBL can be implemented inside any ad hoc routing protocols. It uses Hello messages such as those supplied by OLSR to build its hierarchical structure. However, other ad hoc routing protocol can be used in adding beacon messages to it if such periodic messages are not already provided.

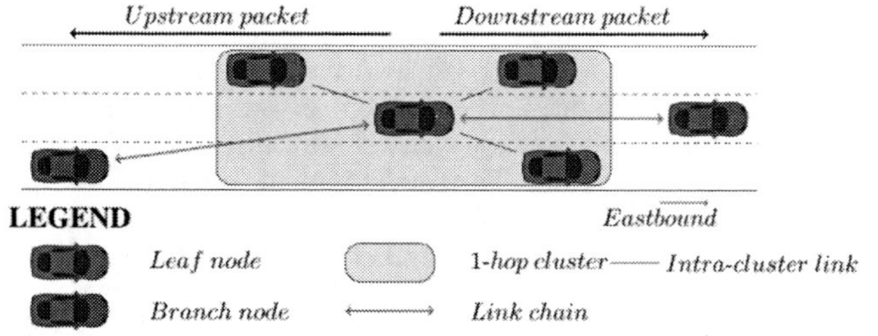

Figure 1. Example of a situation on a three-lane one-way highway

Some definitions are specified as follows:

- **A branch node** (Fig. 1) is a cluster-head node that is elected by other nodes (branch or leaf). It emits HELLO messages like every node, but it is the only one allowed to emit topology control messages (TC), to forward application messages, and to participate in the construction of a chain. In order to control the propagation of a message, based on the application request specified in the header fields, a branch node can forward it to:

- its leaf nodes;
- upstream branch node;
- downstream branch node;
- all branch nodes (including branch nodes of another traffic direction).

Our CBL implementation assumes that these destination options are coded into the link code of the original format of the packets defined in OLSR protocol [24].

- **A leaf node** is an ordinary node which tries to connect itself to the closest branch node. If no branch node is detected, the leaf node elects the neighbor moving with the lowest speed and in the same traffic direction, as a branch. A leaf node sends both HELLO and application messages of which it is the originator.
- **A chain** is a virtual backbone made up of a sequence of branch nodes. Ideally, one chain should be created per traffic direction. On longitudinal road context such as highways, the chains behave as a virtual backbone similar to the one that should be obtained with an infrastructure. It offers to its branch nodes a path to forward application messages over long distance.
- **BranchChoice** is a field added in the HELLO message and containing the address of the elected branch to which the HELLO originator node is connected.
- **The Connection Time (CT)** is the time during which two nodes N_i and N_j could communicate if they kept the same speed. This metric, also called "contact time", has been used in [25]–[27]. CT is approximated using (1). This equation takes into account the positions of the nodes ($[X_i, Y_i]$ for the node N_i and $[X_j, Y_j]$ for the node N_j) their speeds (V_i for the node N_i and V_j for the node N_j), and the maximum radio range (R_{max}):

$$CT = \frac{-(ab+cd) + \sqrt{(a^2+c^2) * R_{max}^2 - (ab-bc)^2}}{a^2+c^2} \qquad (1)$$

$$\left\{ \begin{array}{lcl|lcl} a & = & V_i cos(\sigma_i) - V_j cos(\sigma_j) & b & = & X_i - X_j \\ c & = & V_i sin(\sigma_i) - V_j sin(\sigma_j) & d & = & Y_i - Y_j \end{array} \right.$$

C. HELLO message

Ad hoc network initialization is triggered by the emission of HELLO messages (the beacons) at node level. The Time-To-Live (TTL) value of the HELLO messages is set at 1 to contain its broadcast in the 1-hop neighborhood. At CBL network initialization, there is no chain, no branch, and every node are leaf nodes.

The following information concerning the sender node is added to standard HELLO messages:

- its current location X,Y in Cartesian coordinates (coming from the longitude and latitude in GPS data);
- its current speed V and its steering angle σ, which provides the node direction (extracted from the on-board computer or GPS data for example);
- its current type: leaf or branch node;
- the elected branch node (BranchChoice) to which it is connected;
- the elected upstream branch node to which it is connected (if type=branch) or empty (if type=leaf);
- the elected downstream branch node to which it is connected (if type=branch) or empty (if type=leaf);
- the current list of its 1-hop neighbors with link types;
- the validity time timestamp (Vtime) that indicates the validity period of the information contained in a HELLO message.

D. Local 1-hop neighbor and node internal variables

When processing HELLO messages, each node creates and maintains a local table containing the list of its 1-hop neighbors. We introduce a variable called "Elected" to distinguish in the table whether the neighbor has elected the node either as a branch (the neighbor is a leaf) or as a branch in its chain (the neighbor is a branch). In this case, the variable "Elected" is set at true. Otherwise, it is set at false. This table contains the neighbor addresses and for each neighbor: the link type (UNSPEC if unspecified, LOST if the link is lost, ASYM if the communication link is unidirectional, and SYM if the communication link is bidirectional), the timestamp T1 of the last HELLO messages received from the neighbor, the timestamp T2 when the link became symmetric, its position (X,Y), its speed (V), its direction (σ), and its type (branch or leaf). The table also includes for each neighbor the value of the "Elected" variable.

Each node records and maintains internal variables. These latter are the addresses of the elected upstream and downstream branch nodes, the addresses of BranchChoice, and the timestamp T3. T3 contains the date of a the last HELLO message received from any leaf or branch node that elected it as a relay branch node.

E. CBL scheme

Each reception of a HELLO message by a node N_i and coming from a neighbor node N_j triggers the following procedure (Fig. 2).

1) Algorithm 1. Update the 1-hop neighbor table: CBL scheme uses the same algorithm than the OLSR protocol to update the 1-hop neighbor table and link type. Moreover, it checks for each neighbor in the table if the duration elapsed since the last HELLO message received from it (recorded in

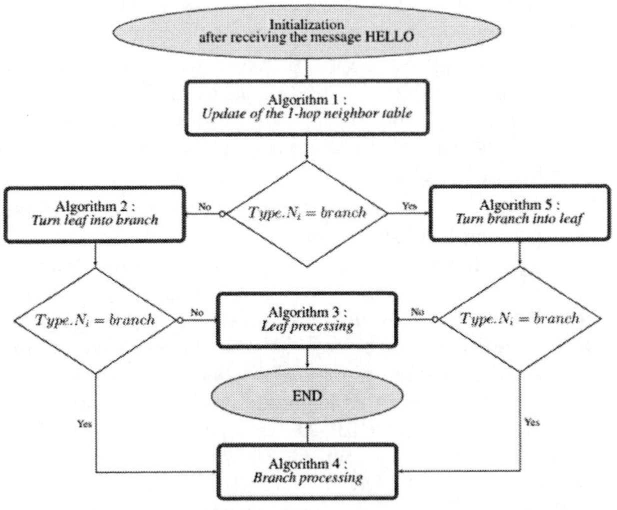

Figure 2. Procedure applied when receiving a HELLO message

timer T1) is higher than Vtime. In that case, the neighbor is removed from the table.

2) Algorithm 2. Turn leaf into branch: when a neighbor N_j has chosen N_i as a branch (the BranchChoice of N_j is set at the address of N_i), the node N_i turns its type into branch. Then, N_i updates its timer T3, initializes its BranchChoice to empty, and writes in its neighbor table that it is elected by N_j. If N_i was already a branch, it will just add automatically the node N_j as its elected upstream or downstream branch nodes to form a chain taking into account its relative position (up- or downstream).

3) Algorithm 3. Leaf processing: a leaf N_i selects a node as its branch, then it writes the branch address in its BranchChoice variable. When several branch candidates are detected, some optimizations are introduced in the choice process. In this work, N_i first looks for branch nodes driving in the same direction, then it chooses the closer according to the distance. Notice that two nodes drive in the same direction when:

$$|\sigma_{Ni} - \sigma_{Nj}| < \sigma_{max} \quad (2)$$

When N_i detects no branch around after a time greater than Vtime with respect to the timer T2, the node sets in its BranchChoice the address of the leaf that is driving in the same direction. Nevertheless, if more than one leaf is a candidate, then the address of the candidate node having the lowest speed is put in its BranchChoice. The chosen leaf will become a branch after receiving a HELLO message from N_i. Selecting branch nodes with low speeds ensures the stability of the chain because, according to [23], the lower the relative speed between the branch nodes, the better the radio communication.

When N_i has N_j address in its BranchChoice variable, if N_j is still a branch node, N_i updates its time counter T3 with the last reception timestamp of the HELLO message from N_j. However, if N_i received no HELLO message from N_j for a time longer than Vtime, it initializes its BranchChoice to empty in order to join a new cluster.

4) Algorithm 4. Branch processing: a branch node N_i participates in the creation of a chain. To this purpose, it elects an upstream branch and a downstream one from its current position, taking into account its trajectory direction. The election process selects the branch nodes that, firstly, currently exists and that, secondly, have not yet joined a chain or have chosen N_i as a branch node of their chain.

By keeping the same upstream and downstream branch nodes for N_i, while N_i location is maintained between them (no overtaking) and N_i is still in their transmission range (it receives their HELLO message), the algorithm favors the stability of the chain.

When N_i detects no branch node either upstream or downstream, it selects among its 1-hop neighbors the leaf node driving in the same direction, provided that this latter brings at least one more link, via its own 1-hop neighborhood, to a new node previously unknown from N_i. Next, if several leaf nodes are found, N_i selects, as BranchChoice, the address of the one that has the highest CT value (see equation 1). This election will change this leaf node into a branch node after the reception of the next HELLO message from node N_i. To select a new branch among leaf nodes, we do not consider a fixed distance threshold to avoid restricting the scheme to only few highway contexts. The advantage of using a communication network metric is to decrease the probability of choosing the closest node, and therefore the risk of frequently breaking the chain when vehicles overtake.

N_i updates the "Elected" variable for N_j in its 1-hop neighbor table: if N_j is connected to N_i, the variable is set at true and N_i updates its timer T3, otherwise the variable is set at false.

The branch node N_i checks for the two selected nodes (upstream and downstream) in its chain the duration elapsed since it received the last HELLO message from this latter. Every neighbor having a duration higher than Vtime is removed.

When the sender N_j is already in the selected nodes (upstream and downstream) in the chain of N_i, this latter checks that the position of N_j (upstream or downstream), its type, and its direction have not evolved, and that N_i address appears in the selected nodes (upstream and downstream) in the chain of N_j. If these conditions are fulfilled, N_i updates the counter T3. If the relative position has changed, it corrects the position of N_j in the or removes it.

5) Algorithm 5. Turn branch into leaf: a branch N_i goes back to leaf type when it received no more HELLO messages from any of the nodes that elected it for a duration above Vtime. This process refers to the value T3.

IV. Performance Evaluation

This section presents the performance evaluation of the proposed CBL scheme over varying highway scenarios and network traffic conditions. SUMO [28] is used in order to generate the mobility traces of the vehicles over three different road networks. The clustering scheme itself is modeled with Matlab.

A. SUMO models

Three different road networks are modeled using the SUMO simulator:

- R1: 5 km-long three-lane one-way highway;

- R2: 5 km-long three-lane two-way highway;

- R3: 5 km-long three-lane two-way highway, an entrance and a highway exit. The exit is located at 1.8 km from the beginning of the section, while the entrance is located at 3 km of it.

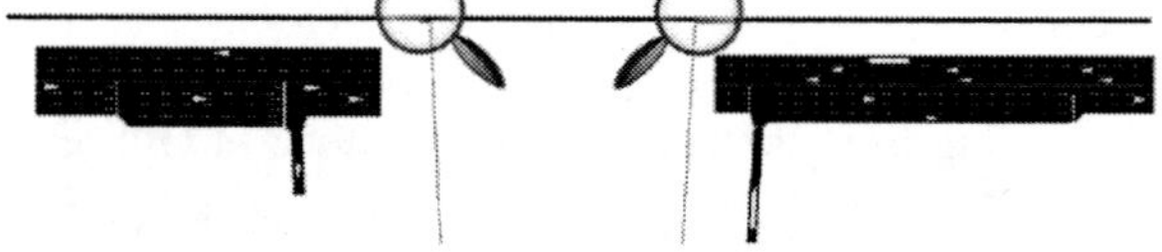

Three different traffic density cases are considered for this evaluation. In each case, a ratio of 1/6 trucks and 5/6 cars is considered. The different traffic densities used are listed in Table I. In the network R3, 25% of the vehicles arrive via the highway entrance, 25% of the vehicles take the exit and 50% of the vehicles just cross the whole road section.

TABLE I. SCENARIOS AND VALUES OF ROAD TRAFFIC DEMAND

Density	Car traffic (veh/h/direction)	Truck traffic (veh/h/direction)	R1	R2	R3
Low	500	100	S1	S4	S7
Medium	2000	400	S2	S5	S8
High	4000	800	S3	S6	S9

Where S1 to S9 are the scenarios.

TABLE II. KINEMATIC PARAMETERS FOR CARS AND TRUCKS

	Acc	Dcc	L	σ	τ	MG	MS	SF	SD
Units	m/s^2	m/s^2	m	-	s	m	km/h	-	-
Cars	2	3	5	0.5	1	2.5	150	1	0.1
Trucks	1	2	15	0.5	1	5	130	0.84	0.1

The default car following model included in SUMO simulator is a variant of the Krauß model: each vehicle drives up to its "desired speed", while maintaining a perfect safety distance with the leader vehicle (*i.e.* the front vehicle). The speed limit is set at 130 km/h, which corresponds to the legal speed limit on highways in France. We define two types of vehicles: cars and trucks. Vehicles are modeled by a set of parameters (the values are given in Table II):

- Acc: the acceleration capability of vehicles;
- Dcc: the deceleration capability of vehicles;
- L: the vehicle length;
- σ: the Krauß driver imperfection (between 0 and 1);
- τ: the driver-desired minimum time headway;
- MinGap (MG): the offset to the leading vehicle when standing in a jam;
- MaxSpeed (MS): the maximum velocity of the vehicle;
- SpeedFactor (SF): the vehicles expected multiplicator for lane speed limits;
- SpeedDev (SD): the standard deviation of the speedFactor;

To achieve realistic car following behavior [29], it is necessary to use speed distributions for the desired speed. Otherwise, if all vehicles have the same desired speed, they will not be able to catch up with their leader vehicle, thus causing unrealistic situation. Therefore, two other parameters have been introduced in order to use speed distributions in SUMO: speedFactor and speedDev. For instance, using $speedFactor = 1$ and $speedDev = 0.1$ will result in a speed distribution where 95% of the vehicles drive at a speed ranging from 80% to 120% of the legal speed limit (Fig. 3).

B. Matlab simulation

Simulation time for each of the nine scenarios is 500 s. Nodes send a HELLO message every 1 s. The thresholds Vtime are set at 3 s. The free space propagation model is used, with a transmission range of 500 m.

C. Performance metrics

Seven performance metrics are considered. The average values reported are picked up when the network is stable (between 200 s and 500 s):

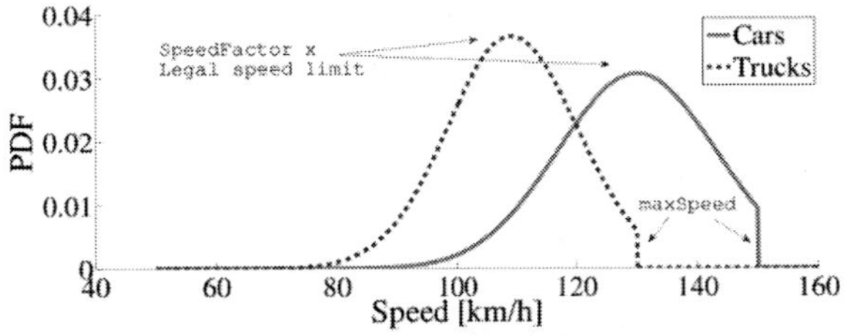

FIGURE 3. PROBABILITY DENSITY FUNCTION OF DESIRED SPEED FOR CARS AND TRUCKS.

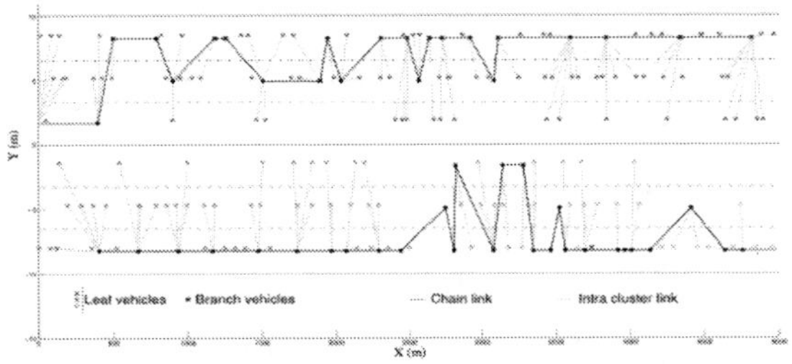

FIGURE 4. STATE OF CHAIN CONSTRUCTION IN THE CASE OF SCENARIO 5 AT TIME T=500S

1) NB_Chains: number of chains in the network
2) Branch/Chain: number of branch nodes per chain
3) 1hop/Branch: number of 1-hop neighbors (in the same traffic direction)
4) Leaf/Branch: number of leaf nodes per branch node
5) Branch_time: duration that a node remains a branch
6) Leaf_time: duration that a leaf node remains attached to the same branch node
7) Leaf/Vanet: percentage of leaf nodes in the network

D. Results

The objective of these evaluations is to analyze the structure created by CBL. Simulating the scenario S5, we observed that CBL leads to two separate chains, one in each road traffic direction (Fig. 4).

Therefore, in this paper, only the results of S2 scenario related to R1 road configuration are commented on, since this latter is the usual configuration of highway traffic in one direction (S2 represents the medium of the three studied densities). However, the results of all the scenarios are in Table III.

About 70% of the time (Fig. 5), there is only one chain as targeted for S2 scenario. Sometimes the chains are broken, mostly due to the changes in the order of the branch nodes inside the chain, but they are quickly reconstructed. The cumulative duration when there are more than one chain is about 30% of the simulation time. It is noticed that chain breaks increase with the density and the road configuration, R3 reaching the highest scores due to a lot of vehicles entering or leaving the road section (see Table III).

When the traffic becomes stable (Fig. 6), after about 150 s of simulation, there are up to 100 vehicles on the highway. We see that up to 75% of the nodes are of leaf type, and only 25% are actually branch nodes. This shows the ability of CBL to optimize the flooding of broadcast traffic since only branch nodes are allowed to relay it upstream, downstream, or both

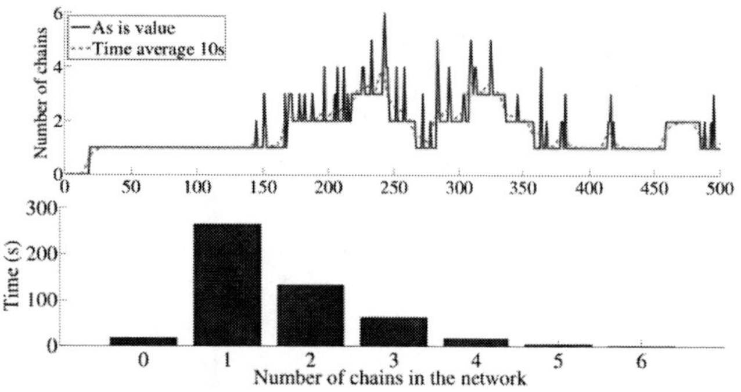

FIGURE 5. NUMBER OF CHAINS AND ACCUMULATED TIME IN THE CASE OF SCENARIO 2

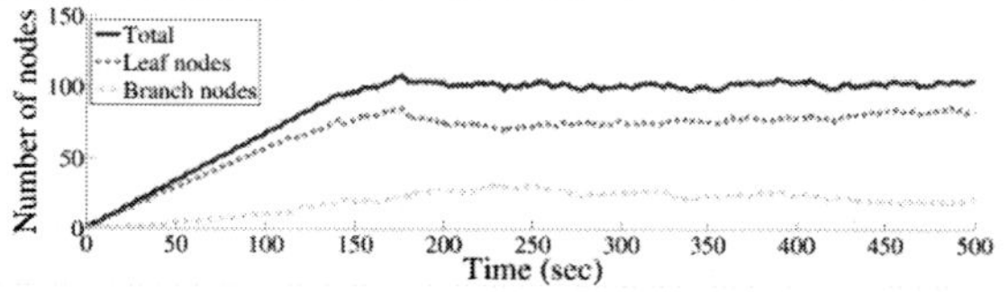

FIGURE 6. NODE TYPE PARTITIONING IN THE CASE OF SCENARIO 2

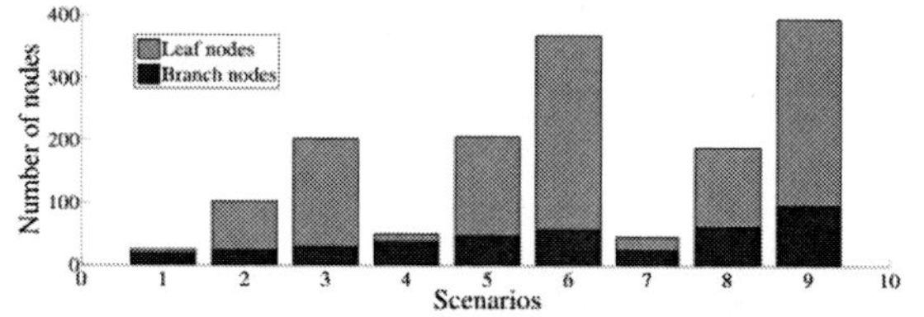

FIGURE 7. NODE TYPE PARTITIONING FOR EVERY SCENARIO (S1 TO S9)

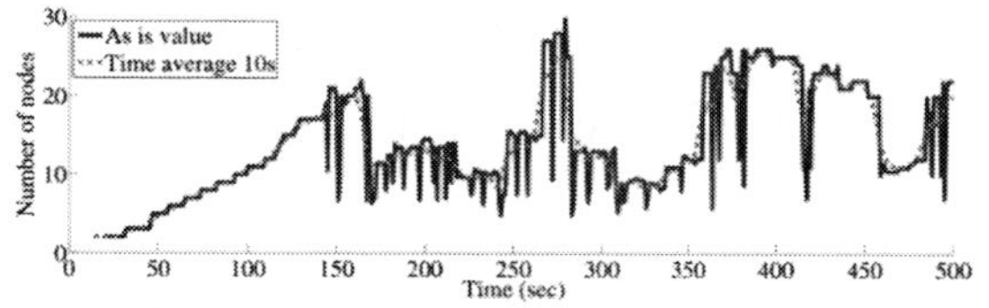

FIGURE 8. NUMBER OF BRANCH NODES PER CHAIN IN THE CASE OF SCENARIO 2

directions according to application requirements. These results are confirmed in every scenario (Fig. 7), except in S1, S4, and S7 where the traffic density is low, and therefore the clustering is less efficient (more than 50% of branch nodes). Intuitively, when the network is sparse, the vehicles are more spaced and there are more isolated nodes that become branch nodes.

Looking at Fig. 8 and 5 together shows that there are about 25 branch nodes when there is only one chain and about 10 branch nodes per chain in the presence of several chains (chain breaks). Each node has an average of 20 1-hop neighbors (Fig. 9). The results show that an average of 5 nodes (25% of the neighbors) choose the same branch node (Fig. 10). A small number of branch nodes are chosen by 75% of their 1-hop neighbors, others by only 10% mainly due to their position (at the end of the chain, *etc.*). CBL parameters may be tuned through time threshold value (Vtime) in order to improve the balance between the number of 1-hop neighbors of a node and that of the leaf nodes choosing it as their branch.

Each selected node remains a branch about 70 s (Fig. 11). Even for a vehicle moving at the lowest speed of 80 km/h, it stays a branch over 1.5 km (3 times the maximum range), which is a significant distance even on a highway. Moreover,

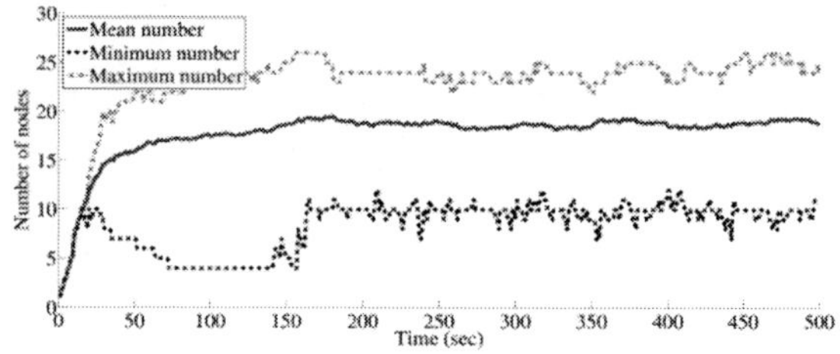

FIGURE 9. NUMBER OF 1-HOP NEIGHBORS IN THE CASE OF SCENARIO 2

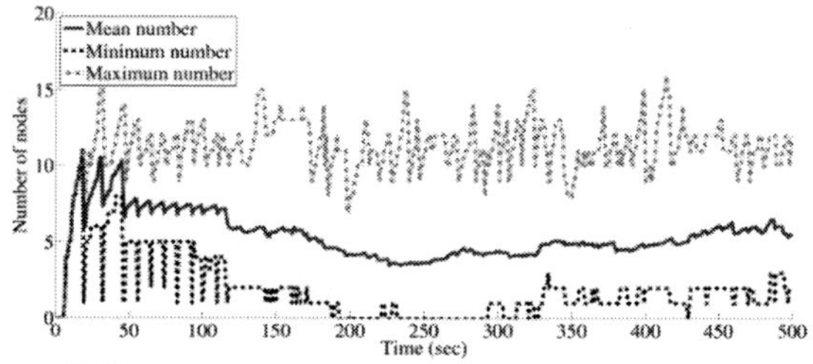

FIGURE 10. NUMBER OF LEAF NODES ATTACHED TO A BRANCH NODE IN THE CASE OF SCENARIO 2

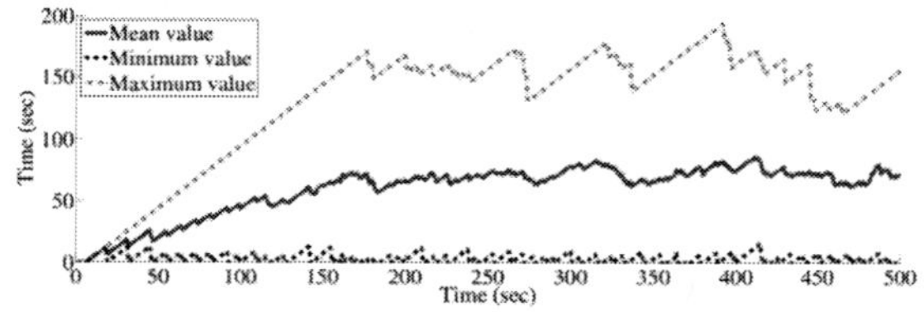

FIGURE 11. DURATION THAT A NODE REMAINS A BRANCH IN THE CASE OF SCENARIO 2

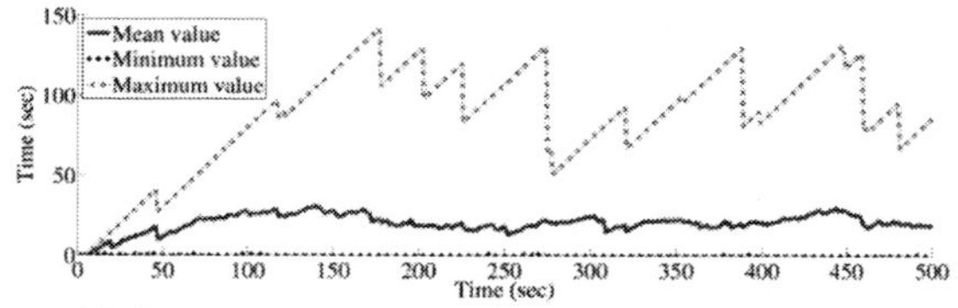

FIGURE 12. DURATION THAT A LEAF REMAINS ATTACHED TO THE SAME BRANCH NODE IN THE CASE OF SCENARIO 2

TABLE III. MEAN RESULT VALUES FOR ALL SCENARIOS

Scenario	1	2	3	4	5	6	7	8	9
NB_Chains	1.65	1.96	2.04	3.29	2.68	3.63	3.36	6.36	9.29
Branch/Chain	13.7	15.9	18.9	12.1	20.4	17.8	9.4	10.4	11.3
1hop/Branch	3.9	18.8	38.6	4.8	21.2	38.2	4.1	17.7	37.8
Leaf/Branch	1.94	4.75	7.03	1.94	4.89	6.84	2.31	3.43	4.42
Branch_time(s)	64.4	71.9	70.0	65.1	73	73.5	59.4	53.6	57.7
Leaf_time(s)	18.8	20.6	21.3	15.6	22.5	22.6	8.6	23.8	16.5
Leaf/Vanet(%)	28	76	85	25	77	84	46	67	76

See section IV-C for the definition of performance metrics.

each leaf node remains attached to the same branch for 20 s on average (Fig. 12). It is also known that most V2V safety applications have a message transmission periodicity ranging from 50 ms to 500 ms. Consequently, that 70-second time represents at least up to 40 alerts from a leaf node relayed by the same branch node to the entire network. These values are almost the same over all 9 scenarios (Table III).

V. Conclusion and Future Work

In the CBL clustering scheme, the vehicles that move at lower speed in the same traffic direction are good candidates (branches) for building a stable backbone that we call a chain. The greater the number of vehicles, the longer the chain. Each vehicle moving faster is a leaf that attaches itself to a branch node covering its current location in order to communicate with the entire VANET. The evaluations show that CBL leads to a structure that may improve VANET performance regarding several metrics. First, the branch nodes represent only 25%, thus allowing optimization of the flooding of broadcast traffic. Then, among all the neighbors of a given branch, only those having the better link quality with this latter (25% to 55%) actually choose it as a branch. The others select other branch nodes, which will result in a global structure with better link quality in the VANET. Finally, this study shows that CBL leads to significant stability since there is only one chain 70% of the time. A node elected as a branch remains a branch for 70 s, and it can serve each of its leaf nodes for 20 s. At 130 km/h, such a leaf would have moved over 720 m while being connected to the same branch node, which is longer than the communication range offered by a fixed RSU. As a clustering approach, CBL can be used in a global VANET architecture including also V2I communications, which makes it compliant to ETSI Geonetworking requirements. However, unlike other geographic routing protocols, CBL is not dependent from any global location service. Future work will consist in finding optimal values of CBL parameters for different traffic conditions, and in comparing with other clustering schemes.

Acknowledgment

The authors acknowledge the support of the CPER ELSAT2020 project which is co-financed by the European Union with the European Regional Development Fund, the French state and the Hauts de France Region Council.

References

[1] F. D. d. Cunha, A. Boukerche, L. Villas, A. C. Viana, and A. A. F. Loureiro, "Data Communication in VANETs: A Survey, Challenges and Applications," INRIA Saclay ; INRIA, report 8498, Mar. 2014, URL: https://hal.inria.fr/hal-00981126/document [retrieved: 2017-06].

[2] R. S. Bali, N. Kumar, and J. J. Rodrigues, "Clustering in vehicular ad hoc networks: Taxonomy, challenges and solutions," *Vehicular Communications*, vol. 1, no. 3, pp. 134–152, Jun. 2014.

[3] S. M. AlMheiri and H. S. AlQamzi, "MANETs and VANETs clustering algorithms: A survey," in *8th IEEE GCC Conference and Exhibition (GCCCE)*. IEEE, Feb. 2015, pp. 1–6.

[4] C. Cooper, D. Franklin, M. Ros, F. Safaei, and M. Abolhasan, "A Comparative Survey of VANET Clustering Techniques," *IEEE Communications Surveys & Tutorials*, pp. 1–1, 2016.

[5] "Intelligent Transport Systems (ITS); Vehicular Communications; GeoNetworking; Part 4: Geographical addressing and forwarding for point-to-point and point-to-multipoint communications; Sub-part 1: Media-Independent Functionality," *European Telecommunications Standards Institute*, no. ETSI EN 302 636-4-1, May 2014.

[6] M. Gerla and J. Tzu-Chieh Tsai, "Multicluster, mobile, multimedia radio network," *Wireless Networks*, vol. 1, no. 3, pp. 255–265, Sep. 1995.

[7] C. Geng, F. Nocetti, J. Gonzalez, and I. Stojmenovic, "Connectivity based k-hop clustering in wireless networks." IEEE Comput. Soc, 2002, pp. 2450–2459.

[8] P. Basu, P. Basu, N. Khan, and T. D. C. Little, "A Mobility Based Metric for Clustering in Mobile Ad Hoc Networks," *International workshop on wireless networks and mobile computing*, pp. 413–418, 2001.

[9] M. Chatterjee, S. K. Das, and D. Turgut, "WCA: A Weighted Clustering Algorithm for Mobile Ad Hoc Networks," *Cluster Computing*, vol. 5, no. 2, pp. 193–204, 2002.

[10] S. Dhurandher and G. Singh, "Weight based adaptive clustering in wireless ad hoc networks," in *International Conference on Personal Wireless Communications*. IEEE, Jan. 2005, pp. 95–100.

[11] E. E. Tsiropoulou, S. T. Paruchuri, and J. S. Baras, "Interest, Energy and Physical-Aware Coalition Formation and Resource Allocation in Smart IoT Applications," in *Information Sciences and Systems (CISS), 2017 51st Annual Conference on*. IEEE, 2017, pp. 1–6.

[12] S. A. Mohammad and C. W. Michele, "Using traffic flow for cluster formation in vehicular ad-hoc networks," in *35th Conference on Local Computer Networks*. IEEE, Oct. 2010, pp. 631–636.

[13] M. N. Avcil and M. Soyturk, "ReSCUE: Relatively Stable Clustering for Unbiased Environments in VANETs." International Wireless Communications and Mobile Computing Conference: IEEE, Aug. 2015, pp. 1049–1055.

[14] R. Mehra, R. S. Bali, and P. Kaur, "Efficient clustering based OLSR routing protocol for VANET," in *Symposium on Colossal Data Analysis and Networking*. IEEE, Sep. 2016, pp. 1–7.

[15] F. Wei, S. Yan, C. Shanzhi, and Z. Longhao, "A mobility metrics based dynamic clustering algorithm for VANETs," in *International Conference on Communication Technology and Application*. IET, 2011, pp. 752–756.

[16] M. Azizian, S. Cherkaoui, and A. S. Hafid, "DCEV: A distributed cluster formation for VANET based on end-to-end realtive mobility," in *International Wireless Communications and Mobile Computing Conference*. IEEE, Sep. 2016, pp. 287–291.

[17] H. R. Arkian, R. E. Atani, A. Pourkhalili, and S. Kamali, "A Stable Clustering Scheme Based on Adaptive Multiple Metric in Vehicular Ad-hoc Networks." *Journal of Information Science & Engineering*, vol. 31, no. 2, pp. 361–386, 2015.

[18] A. Qayyum, L. Viennot, and A. Laouiti, "Multipoint relaying for flooding broadcast messages in mobile wireless networks." IEEE Computer Society, Jan. 2002, pp. 3866 – 3875.

[19] H. Arkian, R. Atani, and S. Kamali, "FcVcA: A fuzzy clustering-based vehicular cloud architecture," in *7th International Workshop on Communication Technologies for Vehicles*. IEEE, 2014, pp. 24–28.

[20] M. Di Felice, L. Bedogni, and L. Bononi, "Dynamic backbone for fast information delivery invehicular ad hoc networks: an evaluation study," in *the 8th ACM Symposium on Performance evaluation of wireless ad hoc, sensor, and ubiquitous networks*. ACM, 2011, pp. 1–8.

[21] C. Wu, S. Ohzahata, and T. Kato, "Can we generate efficient routes by using only beacons? Backbone routing in VANETs," in *24th International Symposium on Personal Indoor and Mobile Radio Communications (PIMRC)*. IEEE, Sep. 2013, pp. 2929–2934.

[22] W. Xiaonan and Q. Huanyan, "Constructing a VANET based on cluster chains," *International Journal of Communication Systems*, vol. 27, no. 11, pp. 2497–2517, 2014.

[23] M. A. Togou, A. Hafid, and P. K. Sahu, "A stable minimum velocity cds-based virtual backbone for vanet in city environment," in *39th Annual Conference on Local Computer Networks*. IEEE, 2014, pp. 510–513.

[24] T. Clausen *et al.*, "Optimized link state routing protocol (OLSR)," *IETF*, no. RFC 3626, 2003.

[25] W. Su, S.-J. Lee, and M. Gerla, "Mobility prediction and routing in ad hoc wireless networks," *International Journal of Network Management*, vol. 11, no. 1, pp. 3–30, 2001.

[26] Y. Li, D. Jin, Z. Wang, L. Zeng, and S. Chen, "Exponential and Power Law Distribution of Contact Duration in Urban Vehicular Ad Hoc Networks," *IEEE Signal Processing Letters*, vol. 20, no. 1, pp. 110–113, Jan. 2013.

[27] I. B. Jemaa, "Multicast Communications for Cooperative Vehicular Systems," phdthesis, Mines ParisTech, 2014.

[28] S. R. Santana, J. J. Sanchez-Medina, and E. Rubio-Royo, "How to Simulate Traffic with SUMO," in *Computer Aided Systems Theory EUROCAST 2015*. Cham: Springer International Publishing, 2015, vol. 9520, pp. 773–778.

[29] L. Rivoirard, M. Wahl, P. Sondi, M. Berbineau, and D. Gruyer, "Using Real-World Car Traffic Dataset in Vehicular Ad Hoc Network Performance Evaluation," *International Journal of Advanced Computer Science and Applications*, vol. 7, no. 12, pp. 390–398, 2016.

Emergency Optimized Low Latency MAC Protocol for VANETs based on VeMAC

Erik Neuhaus
Communication and Networked Systems (ComSys)
Faculty of Computer Science
Westfälische-Wilhelms University Münster, Germany
Email: erikneuhaus@uni-muenster.de

Saleem Raza and Mesut Güneş
Communication and Networked Systems (ComSys)
Faculty of Computer Science
Otto-von-Guericke University Magdeburg, Germany
Email: {saleem.raza, mesut.guenes}@ovgu.de

***Abstract*—Latency is an important metric for time-critical safety applications in vehicular ad-hoc networks (VANETs). Medium access control (MAC) protocol can be greatly exploited to achieve low latency. In this paper, we modify the existing VeMAC protocol frame structure to enhance its latency aspects for time-critical applications. We introduce additional emergency slots for transmission of emergency messages so that vehicles with time-critical emergency messages do not have to wait for their turn for the transmission of such messages. Our modified low latency version of the existing VeMAC protocol shows great improvements in latency for transmission of emergency messages. We analyze the VeMAC protocol and the proposed protocol through simulation and show that the proposed MAC achieves low latency under different scenarios.**

***Keywords*—Vehicular ad-hoc network (VANET); Medium Access Control (MAC); Vehicle-to-Vehicle (V2V); VeMAC.**

I. Introduction

A vehicular ad-hoc network (VANET) is a network of moving vehicles, where the vehicles, equipped with sufficient sensing, computation, and communication capabilities, dynamically form an ad-hoc network without any mandatory infrastructure. VANETs are a special class of mobile ad-hoc networks (MANETs), but having unique characteristics such as high mobility of nodes, dynamic network topology, varying communication environment, varying number of nodes, varying node distribution, etc. VANETs are designed for the purpose to exchange traffic or accidental information between vehicle-to-vehicle (V2V) and vehicle-to-road side unit (V2RSU) networks. VANETs have received tremendous attention due to plethora of applications they support such as *intelligent transportation system* (ITS), traffic information dissemination, infotainment, and the Internet connectivity on the go [1] [2]. Among these, the potential application of VANET is ITS, where the core objective is to control accidents, reduce traffic congestion, and improve driving safety in urban areas.

Owing to importance of VANETs and the multitude of applications supported by the technology, several efforts were taken to standardize it, FCC allocated 75 MHz spectrum in the 5.9 GHz band for Dedicated Short-Range Communication (DSRC) [3] solely for the purpose of V2V and V2RSU communication. DSRC is widely recognized as the IEEE 802.11p Wireless Access in Vehicular Environments (WAVE) and is considered the *de facto* standard for VANETs [4], it is based on IEEE 802.11 MAC and IEEE 802.11a PHY layer [5].

The prime goal of VANETs is to disseminate safety and emergency messages, the timely transmission of such messages is critical to smooth operation of safety applications. As latency is an important performance metric for safety/emergency applications and can be controlled through the medium access control (MAC) layer so this requires for efficient medium sharing. Thus, an efficient MAC protocol should ensure high reliability, low end-to-end latency, and high throughput. Therefore, we analyze and exploit the MAC layer in reducing the latency for safety/emergency messages in the context of V2V communication.

In this paper, we propose the emergency enhanced MAC protocol which is a variant of the VeMAC [6] protocol. VeMAC is a multichannel TDMA MAC protocol which is based on ADHOC MAC [7]. The proposed protocol uses emergency slots to transmit time critical emergency messages in case of road accidens or collisions among vehicles in VANETs. Our proposed protocol achieves low latency for emergency messages under different scenarios and is evaluated through simulation.

The remainder of the paper is organized as follows. In Section II, we discuss VeMAC, its working principle, frame structure and highlight its drawbacks for low latency aspects. Subsequently, Section III gives overview of the desired changes in VeMAC to achieve low latency. In Section IV, we describe the evaluation details of our proposed MAC protocol through simulation. We also discuss different real life scenarios for which the protocol is evaluated. Section V discusses results of the simulation and shows latency improvements through box plots. Finally, Section VI concludes the paper.

II. Related Work

In this section, we review the related work, especially we focus on VeMAC. We thoroughly explain VeMAC and its frame structure.

VeMAC frame structure: VeMAC [6] is a multi-channel TDMA protocol for VANETs, which utilizes two radios. One of the radios is always tuned to the control channel c_0, while the other radio can be tuned to one of the service channels. Each node should acquire exactly one slot on the control channel. The node holds onto this slot until it does not need it anymore or until a merging collision occurs. The collisions occur if two nodes, with the same slot, enter the same two-hop-neighborhood due to their Monument. To reduce the number of collisions, the slots are divided into disjunct sets L, R, and F as shown in Figure 1. The frame structure is split in two disjunct sets based on the general direction of movement of the vehicles. If a node travels in general eastern direction, so $0 - 180^\circ$ degrees of a compass, it would be in the R-subset, the rest in the L-subset as shown in Figure 2. F is an optional set for RSUs which has no direction of movement. That way, vehicles driving in opposing directions are not competing for the same slot and it reduces the relative speed of nodes competing for the same slots and thereby increases the network topology persistence within these sets.

The directions are provided by the GPS unit that each vehicle is mandatory to be equipped with. With the GPS unit it is possible to synchronize the frames through the pulse per second (PPS) signal provided by each GPS receiver. A frame should start at the beginning of each GPS second.

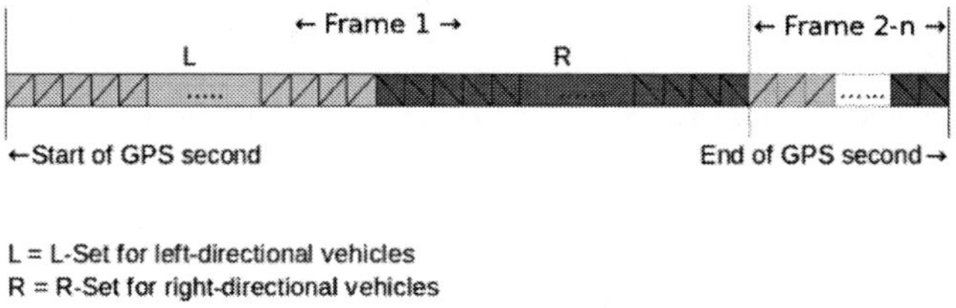

Figure 1. Frame structure of VeMAC [6].

The VeMAC protocol proposes a time division in a periodical frame structure of fixed duration. One frame consists of 100 slots, where the length of one slot is of 1 ms duration, hence a frame length of 100 ms. Each node should transmit periodically one message per frame in its allocated slot. The message consists of a header field, two fields to organize the allocation of slots on the service channels as well as one field for exchange of information for high-priority short applications.

Each node should have a unique random ID to identify the node. The header of the message of node x includes, amongst others, the set $N(x)$ which is the set of IDs of the one-hop neighbors of node x on channel c_0, from which node x has received packets on channel c_0 [6] in the previous 100 slots. With the sets $N(y)$ of each one-hop neighbor y, the node is able to determine which slots are used by its two-hop neighborhood. These slots, that the node must not use in the next 100 time slots, are denoted by $T_0(x)$. With this information, the node builds the set of available slots $A(x) = \overline{T_0(x)}$ respectively with regard to the directional division, e.g., $A(x) = \overline{T_0(x)} \cap R$ for vehicles driving in eastern direction. With the provided information the node is able to solve the hidden-terminal problem.

Node x also determines whether or not all of its one-hop neighbors received its last broadcast by looking for its ID in the right slot in all $N(y)$. It thereby constitutes a reliable broadcast mechanism. Due to the regular transmitting, there exists an upper bound for transmission of messages of 100 ms. However 100 ms is a long time in high mobility scenarios.

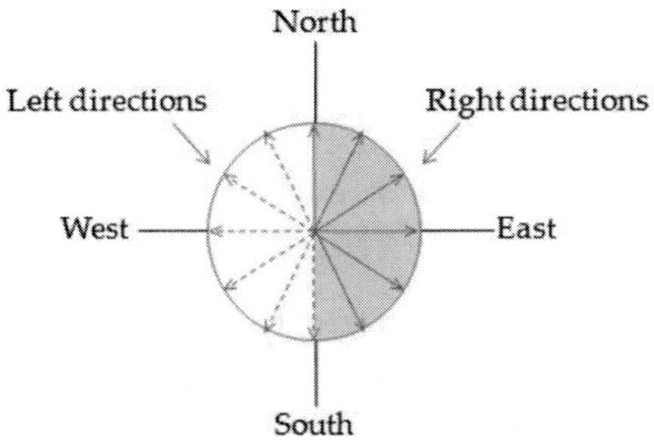

Figure 2. Division of node per direction [6] showing the distinction for the L set and R set.

Limitation of VeMAC for emergency messages: In 100 ms, a car traveling on the highway with the recommended speed of 130 km/h already covers a distance of 3.6 meter and many cars drive considerably faster on the highway in Germany. While the 100 ms limit should be sufficient in normal use, it might be too long for emergency situations where fast responses are crucial.

III. Proposed Low-Latency Optimized MAC

To reduce the latency in emergency situations, in this paper, we propose emergency enhanced VeMAC (EEVeMac), which is variant of the VeMAC protocol, by introducing emergency slots at the beginning of the L set in slot 0 and R set in slot 50 as shown in Figure 3. They are evenly distributed across the frame structure to reduce the average distance to any other slot. The slots are based on the principle of CSMA for the transmission of time-critical emergency data. In case of an emergency, a vehicle wants to send time-critical data to notify other vehicles of its situation. In this way, instead of waiting for its next allocated slot, the vehicle can use these additional emergency slots to quickly transmit the messages and avoid catastrophic situations. With additional slots, vehicles have three possible slots instead of one to transmit their data during emergency situations, effectively bringing down the upper bound latency to 50 ms. While the upper bound latency is 50 ms, the median average is further reduced since a slot is able to choose from three possible slots for emergency transmission instead of one.

While the original VeMAC protocol does not define the exact nature of $N(x)$ for node x, we implemented them in both VeMAC and EEVeMAC as pair of ID and slot number to preserve the reliable broadcast mechanism in EEVeMAC. Through this modification, an ID can be twice in a set. A receiving node then thereby acknowledges the reception of an emergency message by including the ID of the sending node in the emergency slot number in which it received the emergency message. This implementation decision will extend the length

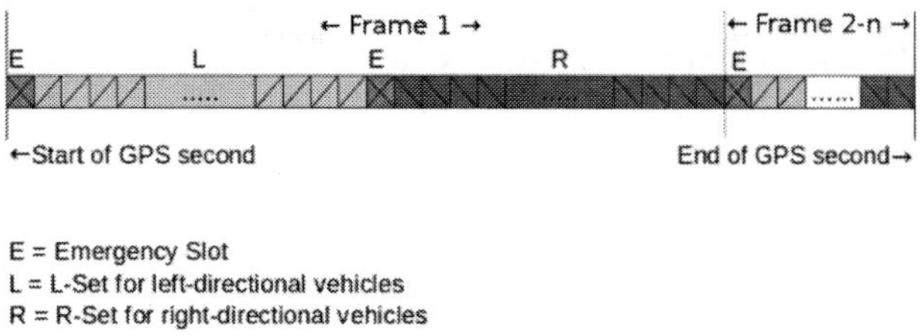

Figure 3. Frame structure of VeMAC with emergency slots. The emergency slots are set at the beginning of the L set respectively the R set.

of the regular message by a maximum of 100 bytes (88 bits total, 7 bits for representation of numbers up to 128, rounded up to 8, multiplied by 100 slots).

IV. Performance Evaluation

We evaluate the proposed protocol in OMNeT++ [8] simulation environment together with Veins [9] and SUMO [10]. Veins is an open source simulation framework for vehicular network simulation. It bi-directionally couples two softwares: OMNeT++ is utilized for network simulation and the open source traffic suite SUMO of the German Aerospace Center provides the traffic simulation data. SUMO has several car-following-models and lane-changing-models to reproduce realistic traffic behavior. Veins integrates MiXiM [11] for modeling physical layer effects and provides realistic interference models. For our simulation we use the two-ray-interference model provided by Veins [12].

Scenario: The highway interchange Münster south, Germany was created in SUMO as shown in Figure 4 and provided with traffic statistic of the state office for road construction NRW [13] to achieve a realistic traffic scenario.

Two scenarios "straight" and "interchange" with reduced road traffic and normal road traffic were tested to examine the influence of node numbers on collisions. In the straight scenario, only traffic from northern and southern directions was present; in the interchange scenario vehicles started from each direction. In each scenario, 20% of cars were presumed to change from one highway to the other highway with 10% in each direction of the highway. The road traffic was implemented with the traffic flow functionality of SUMO which regularly introduces vehicles based on the number of vehicles per hour. The scenario consists of a car that drives on the highway in northern direction and wants to change the highway in western direction on the interchange. It breaks down on the clover interchange lane and sends an emergency message. The car drove in north-west direction and hence it has a regular slot in the first half of the frame structure. In each scenario, the emergency was set to three different slots. To slot 1, directly after an emergency slot, to slot 25, in the middle between two emergency slots, and to slot 49, right before an emergency slot. Each configuration was run with 50 repetitions to achieve a good confidence interval. In combination with the other parameters as summarized in TABLE I, this resulted in 600 simulation runs.

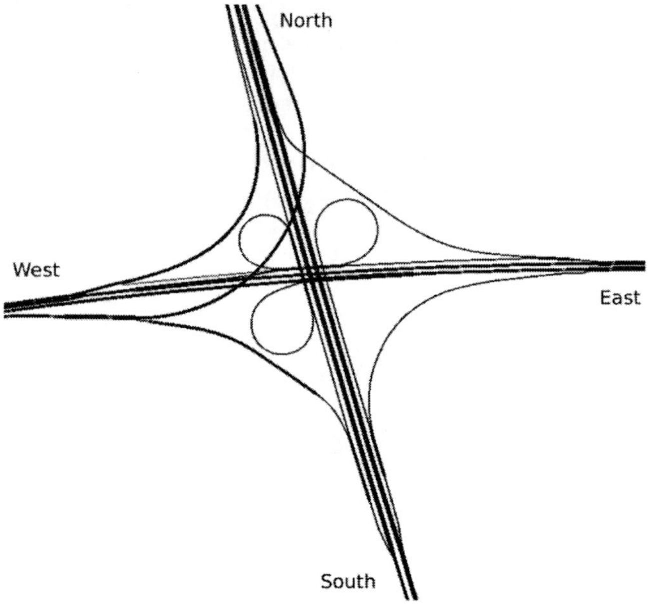

Figure 4. Interchange Münster south, the car with the emergency tries to travel from south direction to west direction and breaks down in the clover interchange line.

TABLE I
Overview of simulation parameters in the two scenarios.

Parameters	Value	
Scenario	Straight	Interchange
Traffic flow from direction	North/South	North/East/South/West
vehsPerHour (total value)	4645	9476
Use of Emergencyslots	False/True	False/True
Emergency in Slot	1/25/49	1/25/49
Replications	50	50
Simulation duration	80 sec.	80 sec.

In addition to the aforementioned scenarios, we conducted a scenario "dense traffic" with additional cars to simulate extremely dense traffic as it would be expected in urban traffic. We conducted it with the same parameters as the "Interchange" scenario, but increased the numbers of vehicles to 13600 vehicles per hour.

V. Results

For the evaluation of EEVeMAC protocol, we measured two values. The latency from the moment the emergency occurred to the moment the one-hop neighbors receiving the emergency message. The second evaluation value consists of the occurrence of collisions, which were calculated to the arithmetic average per node. The results showed an overall improvement of the latency as further explained below.

A. Latency in straight scenario

In the straight scenario, there were 21 nodes in transmitting range of the emergency vehicle at the moment of the emergency situation. The emergency message took a median

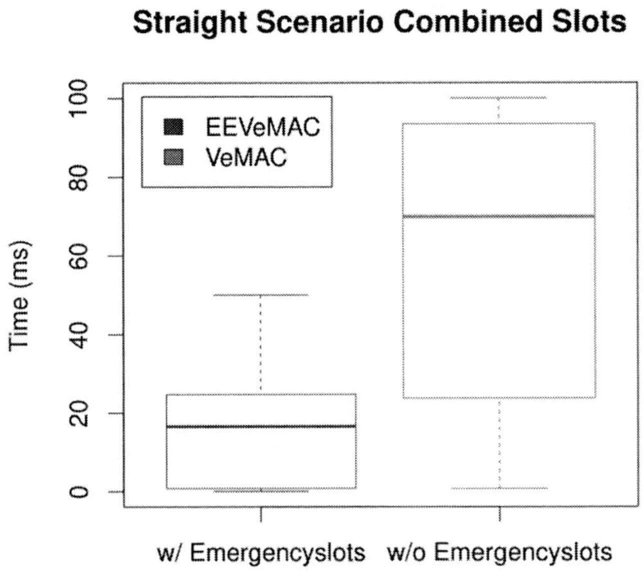

Figure 5. Evaluation results of straight scenario: On the left the latency results of EEVeMAC, on the right the latency results of original VeMAC.

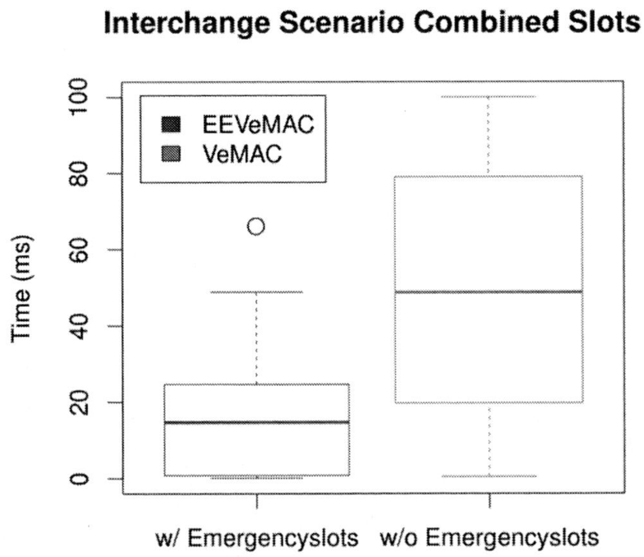

Figure 6. Evaluation results of interchange scenario: On the left the latency results of the EEVeMAC, on the right the latency results of original VeMAC.

time of 69.99 ms to reach the one-hop neighbors of the emergency vehicle in the original VeMAC. With EEVeMAC, with the addition of emergency slots, this value was reduced to 16.57 ms as shown in Figure 5. If the emergency occurred in the first slot after an emergency slot, the median latency was closest to the original protocol with 34.58 ms (VeMAC) vs. 25.01 ms (EEVeMAC) since there is a good chance that the regular slot of the emergency vehicle is between the slot in which the emergency occurs and the next emergency slot. If there is a regular slot in between the emergency and an emergency slot, there is no difference between both protocols as they would both transmit the emergency message in the regular slot. The improvement occurs in the cases where the emergency slot is used. The biggest improvement could be measured with the emergency in slot 49, directly in front of an emergency slot with 73.79 ms (VeMAC) vs. 0.61 ms (EEVeMAC) as shown in Figure 8 (a). Without the emergency slots, the emergency vehicle has to wait at least 50 ms if it does not have slot 49 as its regular slot. It can not transmit in the slot numbers 50-99 since the emergency vehicle is driving in north western direction and hence prefers a slot in the L-set in slot numbers 0-49 of the frame. With the emergency right between two emergency slots, the median latency was improved by 57.61 ms from 81.730 ms in the original VeMAC to 24.120 ms in EEVeMAC with emergency slots.

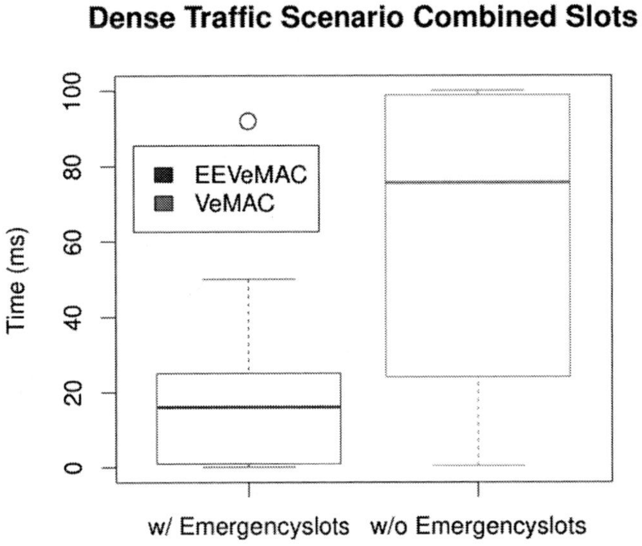

Figure 7. Evaluation results of dense traffic scenario: On the left the latency results of the EEVeMAC, on the right the latency results of original VeMAC.

B. Latency in interchange scenario

The results of the interchange scenario with traffic flow from each direction showed similar improvements as shown in Figure 6. In this scenario, 35 nodes were in transmitting range present during the emergency situation. The median latency was improved by factor 3 from 48.73 ms (VeMAC) to 14.66 ms (EEVeMAC). The biggest improvement could be once again measured if the emergency occurred in the slot right before an emergency slot 75.61 ms in the original VeMAC vs. 0.61 ms in the EEVeMAC, the smallest improvement with the emergency right behind an emergency slot 27.45 ms vs. 24.7 ms. When the emergency occurred in the middle between two emergency slots, the median latency still shows an improvement of 14.6 ms with 38.67 ms measured in the VeMAC and 24.07 ms in the EEVeMAC as shown in Figure 8 (b).

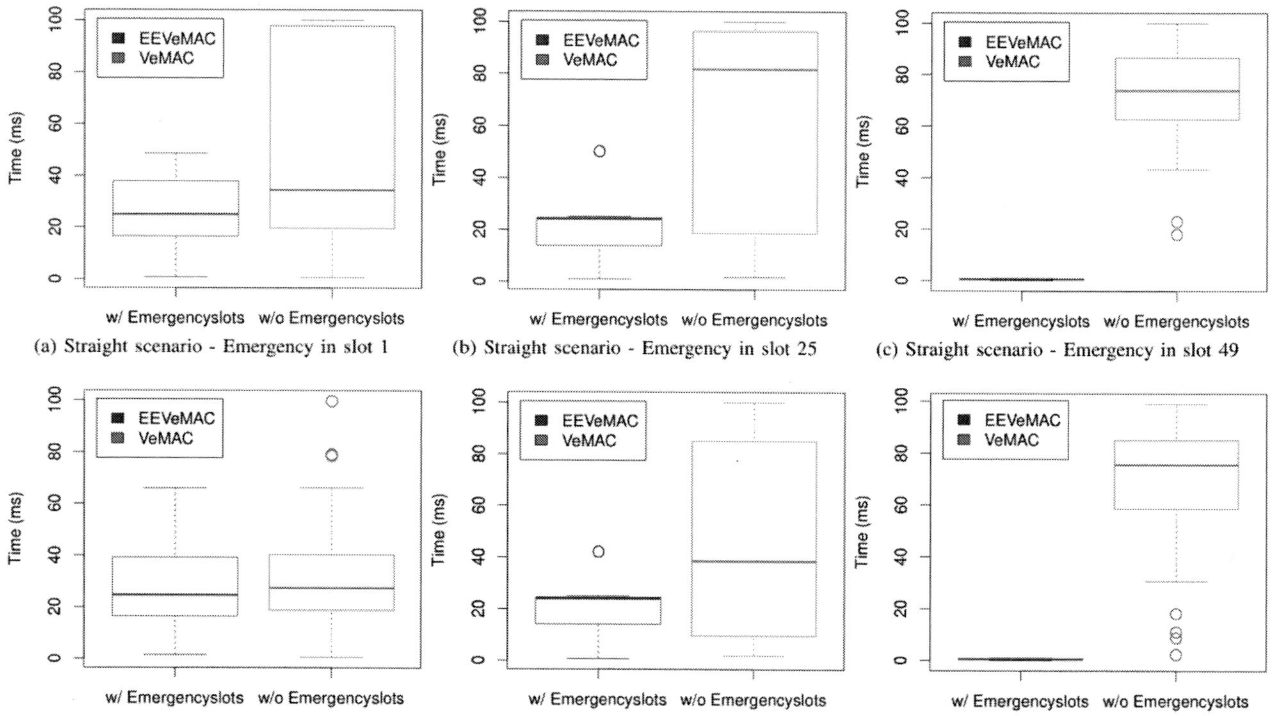

(a) Straight scenario - Emergency in slot 1 (b) Straight scenario - Emergency in slot 25 (c) Straight scenario - Emergency in slot 49

(d) Interchange scenario - Emergency in slot 1 (e) Interchange scenario - Emergency in slot 25 (f) Interchange scenario - Emergency in slot 49

Figure 8. Overview of evaluation of latency results for the straight scenario and the interchange scenario with the emergency in slots 1, 25, and 49.

C. Latency in dense traffic scenario

In this scenario with additional traffic, 50 nodes were in range of the emergency vehicle. The results showed overall similar results as in the normal interchange scenario. The median latency was measured slightly higher with 16.05 ms (EEVeMAC) vs. 75.65 ms (VeMAC) as shown in Figure 7.

D. Collisions

The reservation of two slots for transmission of emergency messages results in a higher expectation of collisions. Instead of 100 slots for transmission of their regular message, the nodes only have a maximum of 98 slots to choose from. Therefore, we also measured the number of collisions. As a measurement, we took the average number of collisions per node. The number of collisions increases in the straight scenario from 0.04575 average collisions per node in the original VeMAC to 0.04747 average collisions per node in the EEVeMAC. The results of the second scenario show that the effect is negligible compared to the effect the numbers of nodes have. The VeMAC had 0.29483 collisions per node whereas the EEVeMAC had 0.29096 collisions per node on average. The average number of collisions increases further with additional traffic in the dense traffic scenario. In the simulation runs with VeMAC, 0.52835 collisions occured whereas EEVeMAC measured 0.66370 collisions.

VI. Conclusion

The introduction of emergency slots in VeMAC shows great improvements for the transmission of high-priority emergency messages. Instead of median latencies of up to 80+ ms we achieved in our test configurations a maximum of median latencies smaller than 25 ms. The latencies were reduced by factor of 3-4. The median and average latencies were improved in each study configuration. The reduction of available slots for regular transmission through the reservation of emergency slots had negligible effects on the rate of collisions.

Further, in situations where several vehicles try to send out an emergency message at the same time, competition emerges and latency increases as the vehicles fail to acquire the emergency slot. The vehicles can still use their normal slots to transmit the emergency message which means that the average latency converges to the maximum latency of VeMAC, e.g. 100 ms. The performance of EEVeMAC in emergency situations with two or more involved vehicles remains to be evaluated. Moreover further research is to be conducted in regards to the optimal number of emergency slots and their effect on collision rates.

References

[1] P. Papadimitratos, A. D. L. Fortelle, K. Evenssen, R. Brignolo, and S. Cosenza, "Vehicular communication systems: Enabling technologies, applications, and future outlook on intelligent transportation," *IEEE Communications Magazine*, vol. 47, no. 11, pp. 84–95, November 2009.

[2] M. Kihl, *Vehicular Network Applications and Services*. CRC Press, 2009.
[3] J. B. Kenney, "Dedicated short-range communications (dsrc) standards in the united states," *Proceedings of the IEEE*, vol. 99, no. 7, pp. 1162–1182, July 2011.
[4] "Ieee standard for information technology– local and metropolitan area networks– specific requirements– part 11: Wireless lan medium access control (mac) and physical layer (phy) specifications amendment 6: Wireless access in vehicular environments," *IEEE Std 802.11p-2010*, pp. 1–51, July 2010.
[5] D. Jiang and L. Delgrossi, "Ieee 802.11 p: Towards an international standard for wireless access in vehicular environments," in *Vehicular Technology Conference, 2008. VTC Spring 2008. IEEE*. IEEE, 2008, pp. 2036–2040.
[6] H. A. Omar, W. Zhuang, and L. Li, "VeMAC: A TDMA-Based MAC Protocol for Reliable Broadcast in VANETs," *IEEE Transactions on Mobile Computing*, vol. 12, no. 9, September 2013.
[7] F. Borgonovo, A. Capone, M. Cesana, and L. Fratta, "Adhoc mac: New mac architecture for ad hoc networks providing efficient and reliable point-to-point and broadcast services," *Wireless Networks*, vol. 10, no. 4, pp. 359–366, 2004.
[8] A. Varga and R. Hornig, "An overview of the omnet++ simulation environment," in *Proc. of the 1st international conference on SIMTOOLS*, 2008, p. 60.
[9] Date last visited 2017.03.20. [Online]. Available: http://veins.car2x.org
[10] Date last visited 2017.03.14. [Online]. Available: http://sumo.dlr.de
[11] A. Köpke *et al.*, "Simulating wireless and mobile networks in omnet++ the mixim vision," in *Proceedings of the 1st international conference on SIMTOOLS*, 2008, p. 71.
[12] C. Sommer, S. Joerer, and F. Dressler, "On the applicability of two-ray path loss models for vehicular network simulation," in *4th IEEE Vehicular Networking Conference (VNC 2012)*. Seoul, Korea: IEEE, November 2012, pp. 64–69.
[13] Date last visited 2017.03.23. [Online]. Available: https://www.strassen.nrw.de/

Vehicle Identification Based on Secondary Vehicle Identifier - Analysis, and Measurements -

Markus Ullmann* † Tobias Franz,† and Gerd Nolden*
* Federal Office for Information Security
D-53133 Bonn, Germany
Email: {markus.ullmann, gerd.nolden}@bsi.bund.de
† University of Applied Sciences Bonn-Rhine-Sieg
Institute for Security Research
D-53757 Sankt Augustin, Germany
Email: markus.ullmann@h-brs.de, tobiasfranz91@gmail.com

***Abstract*—Increasingly, vehicles will be equipped with information and communication technologies, e.g., wireless communication technologies like IEEE 802.11x (Wi-Fi), Bluetooth, mobile communication, etc. These communication technologies enable identification based on identifier used in communication protocols. Today, the Vehicle Identification Number, and the license plate are regarded as vehicle identifier. With new communication technologies used in modern vehicles secondary vehicle identifier comes up. This paper analyzes identification of vehicles based on wireless communication interfaces and presents results of first real measurements of vehicular Bluetooth and Wi-Fi interfaces.**

***Keywords*–Vehicle Identification; Vehicle Identifier; Wireless Vehicle Interfaces; Privacy; Vehicle Tracking**

I. Introduction

The IT architecture of vehicles has significantly changed during the last 10 years. This is shown by the increasingly availability of components for driving assistance: lane keeping support, traffic jam assist, automatic parking assistant, remote parking assistant and so on. This is a prestage of automatic driving, which is one of the main challenges in automotive engineering at the moment. Besides driving assistance, modern vehicles are equipped with wireless interfaces, e.g., Bluetooth to connect devices (smart phones, tablets, etc.) to the multimedia component (head-unit) of the vehicle. In addition, head-units are more and more able to establish a Wi-Fi hot spot to support internet access for vehicle passengers. Furthermore, the vehicle-2-vehicle communication technology (V2V) will be deployed based on IEEE 802.11p technology in the near future. V2V is one feature of Intelligent Transport Systems (ITS).

Today, only the Vehicle Identification Number (VIN), and the license plate are regarded and used as official vehicle identifier. This paper analyses vehicle identification capabilities of wireless communication interfaces, first, which can be used for vehicle identification and tracking. Next, first results of enforced measurements of vehicular Bluetooth interfaces and vehicular Wi-Fi hotspots are presented. The communication interfaces are built in the vehicle to support communication services for occupants. But we show that these services are available outside the vehicle and can be misused for unauthorized identification and tracking. We only use cheap measurement equipment (partially open source), which is publicly available. Especially, the applied smart phone measurement apps can be used by everyone with every modern Android compatible device for identification of vehicles based on Bluetooth. The analysis of Secondary Vehicle Identifier based on Bluetooth and Wi-Fi features is quite new and published here, first.

The subsequent sections of this paper are organized as follows: Section II is a description of related work. Subsequently, identifiers for ITS vehicle stations are presented in Section III. Section IV describes implemented wireless technologies in modern vehicles and analyzes identification capabilities. Aim of the performed tests, used test equipment and investigated test vehicles are presented in Section V. Results of real measurements of Bluetooth and Wi-Fi identifier are given in Section VI. Finally, we summarize our results, and mention open research issues.

II. Related Work

A classification of vehicle identifier is given in [1], which is applied in this paper, too. Hwajeng et al. supposed a vehicle identification and tracking system based on optical vehicle plate number recognition [2]. Tracking of devices based on Bluetooth interfaces is already discussed for a lot of applications, e.g., indoor localization [3] or wireless indoor tracking [4]. But only in [5], an analysis in Jacksonville, Florida, to capture vehicle traffic streams is described. Therefore, a set of Bluetooth receivers was located at the roadside on specific streets to capture the Bluetooth MAC ID of crossing vehicles. Besides Bluetooth, IEEE 802.11 compliant devices were suggested for real-time location tracking in indoor and outdoor environments [6].

Since the 1th of November 2014, vehicles and motorhomes have to be equipped with a Tire Pressure Monitoring System (TPMS) within Europe. They can be separated in direct and indirect TPMS. Direct TPMS means that specific physical sensors measure the air pressure of the tires. These sensors communicate wireless with the vehicle and transmit an identifier of 28 to 32 bit length. There are different wireless technologies available for 125 kHz, 315 kHz, and 433 MHz.

A detection range of up to 40 m for direct TPMS is mentioned in [7].

Besides the identification of vehicles based on static identifiers used in communication protocols different feature based identification methods are suggested. One approach is the identification of vehicles based on noise features (individual noise spectrum) [8].

Further identification techniques allow wireless devices to be identified by unique characteristics of their analog (radio) circuitry; this type of identification is also referred to as physical-layer device identification. It is possible due to hardware imperfections in the analog circuitry of transmitter introduced at the manufacturing process. A good overview concerning the physical fingerprinting of different wireless communication technologies is given in [9].

III. ITS Vehicle Identifier

Here, we categorize the available identifiers of vehicles into two classes. Primary vehicle identifier represent such identifiers which will be typically regarded today, e.g., the Vehicle Identification Number (VIN). Secondary Vehicle Identifier come up with new information technologies used in modern vehicles.

A. Primary Vehicle Identifier

To date, each vehicle is identifiable based on the distinct VIN. In some areas, the VIN is integrated as human readable information in the windscreen of vehicles.

Besides the VIN, vehicles are marked with a licence plate, which is already used for identification.

With the deployment of the V2V technology vehicles will be equipped with a long term ECC key pair and an appropriate certificate [10] [11]. This certificate will become an additional primary vehicle identifier.

B. Secondary Vehicle Identifier

Modern vehicles are equipped with multi-media components (head-unit), which are able to establish communications with electronic devices of drivers or passengers. Typically, wireless communication technologies, e.g., Bluetooth, are used for that purpose.

A Bluetooth multi-media device emits a static 48 bit MAC identifier. The MAC ID is composed of two parts: the first half is assigned to the manufacturer of the device, and the second half is assigned to the specific device. In addition, each Bluetooth device emits a "User-friendly-name" which is typically alterable. Bluetooth devices operate in the ISM band (2.4 to 2.485 GHz).

Moreover, vehicle head-units allow any Wi-Fi equipped laptop, tablet or mobile phone to access the internet within the ITS vehicle station while travelling if the head-unit has mobile communications capabilities. But head-units configured as access point need a unique Service Set Identifier (SSID) or network name to connect devices. In addition, each head-unit needs an unique MAC address.

If vehicles are equipped with mobile communication capabilities an International Mobile Subscriber Identity (IMSI) is required. That is an unique ID to identify a mobile device within the network. In addition, a SIM card with an assigned mobile phone number is needed for mobile communication.

In [9], physical fingerprinting of wireless transmitter is investigated. So, a complete feature set for physical fingerprinting of a transmitter is a secondary vehicle identifier. So far mentioned vehicle identifiers are sufficient for identification all the time. Furthermore, vehicle identifier with a limited validity period, e.g., pseudonymous certificates (termed authorization tickets by ETSI) exist. Pseudonymous certificates come up with the V2V technology.

Initially, secondary vehicle identifier have no formal character in contrast to a licence plate or VIN. But it is technically very easy to capture Bluetooth and Wi-Fi identifiers of a vehicle shown in Section VI. So, attackers can misuse them for their purposes.

IV. Wireless Technologies

Here in this section wireless technologies, which are applied in vehicles are described, first. In addition an analysis concerning identification capabilities based on wireless communication technologies is given. We only address local wireless communication technologies, which are quite easy to detect and omit mobile communications like GSM, LTE, or 5G.

A. Bluetooth

The concept behind Bluetooth is to provide a universal short-range wireless communication capability using the 2.4 GHz band, available globally for unlicensed low-power uses. Bluetooth is specified by the Bluetooth special interest group [12].

1) Technology: Bluetooth provides support for three application areas using short-range wireless connectivity:

- Data and voice access points: Bluetooth facilitates real-time voice and data transmissions by providing effortless wireless connection of portable and stationary communications devices
- Cable replacement: Bluetooth eliminates the need for numerous, often proprietary cable attachments for connection of practically any kind of communication devices. The range of each radio depends on the output power (up to 100 m)
- Ad hoc networking: A device equipped with a Bluetooth radio can establish instant connection to another Bluetooth radio as soon as it comes into range

In vehicles Bluetooth is used for connecting a smart phone to the:

- Hands-free phone system
- Vehicular head-unit to use the loudspeaker of the head-unit to output the music from the smart phone

The Bluetooth architecture is divided into different layers. It starts with the Radio Frequency (RF) Layer, also termed physical layer (PHY). To be resistant to disturbance a frequency hopping spread spectrum (FHSS) is used. Bluetooth devices operate in the ISM band (2.4 to 2.485 GHz). This frequency range is divided into channels of a bandwidth of 1 MHz. There are 79 useable channels. Three classes of transceivers are available with different output power: 1 mW, 2,5 mW and 100 mW.

At first, Bluetooth devices have to establish a connection, termed pairing, to exchange data. This procedure is initiated by the master device based on the inquiry process. During this process Bluetooth devices respond with inquiry reply messages including BD_ADDR and clock rate (CLK), etc. During the pairing process the jump sequence for sharing the channels is calculated by the master device and synchronized with the slave devices.

There exist a range of Bluetooth versions from Bluetooth 1.0a (published 1999) to Bluetooth 5.0 (published 2016).

2) Identification Capabilities: A Bluetooth multi-media device emits a static 48 bit MAC identifier (BD_ADDR). The MAC ID is composed of three parts: Lower Address Part (LAP), Upper Address Part (UAP), and Nonsignificant Address Part (NAP). LAP (24 bit) and UAP (8 bit) are assigned to the manufacturer of the device, and NAP (16 bit) is assigned to the specific device. In addition, each Bluetooth device emits a "User-friendly-name" which is typically alterable. BD_ADDR and the "User-friendly-name" are the primary identifier. In addition, the data set of a Bluetooth device: CLK, Bluetooth device profile, and the Host Controller Interface (HCI) can be used for identification purposes, too (Table I).

B. *Wireless Local Area Network (Wi-Fi)*

Primary, Wi-Fi is based on the communication standards which was made for cable based Local Area Networks (LAN), IEEE 802.11 X.

1) Technology: Briefly spoken, Wi-Fi devices support two different modes:

- Ad-Hoc mode, termed independent BSS (IBSS): Wi-Fi devices communicate peer-to-peer. During the communication data pakets are send to all devices of the network but discarded by the devices if the destination address does not fit
- Access point mode, termed Basic Service Set (BSS): All Wi-Fi devices are connected with the access point (hot spot)

Head-units of modern vehicles provide Wi-Fi hot spots. So any Wi-Fi equipped laptop, tablet or mobile phone is able to access the internet within the vehicle while travelling if the head-unit has mobile communication facilities (GSM, LTE).

Different Wi-Fi Standards exist: IEEE 802.11b / g / a / n / ac. They differ in the used frequency band (2,4 GHz and/or 5 GHz), and communication speed (1 Mbit/s ... 6,96 Gbit/s). The frequency band is splitted into channels (2,4 GHz: 13 channels with a bandwidth of 5 MHz, whereby 5 channels are needed to establish a network). In the 5 GHz frequency band a 455 MHz frequency bandwidth is reserved for Wi-Fi to establish 18 different Wi-Fi networks.

One of the management frames in IEEE 802.11 based Wi-Fis are beacon frames. Beacon frames are transmitted periodically to announce the presence of a wireless LAN and contain information about the network. Beacon frames are transmitted by the access point in an infrastructure basic service set (BSS). In IBSS network beacon generation is distributed among the stations.

2) Identification Capabilities: Primary identifier are:

- Basic Service Set ID (BSSID) or MAC address of the Wi-Fi device and
- SSID (primary name associated with an 802.11 wireless local area network with a maximum length of 32 characters)

TABLE I. TECHNOLOGY SPECIFIC IDENTIFICATION FEATURES

Technology	First Level Features	Second Level Features
Bluetooth	MAC ID (BD_ADDR) "friendly name"	CLK, Bluetooth device profil Host Controller Interface
IEEE 802.11 X (Wi-Fi)	MAC ID (BSSID) "SSID"	Information in Beacon Frames

In addition, information in Wi-Fi beacon frames could be used for identification, too (Table I).

V. MEASUREMENTS

In this section the performed test cases and the used test equipment is described.

A. *Aim of the Measurements*

With the measurements we investigate vehicular Bluetooth as well as Wi-Fi communication capabilities especially for identification purposes outside the vehicle. Therefore, following measurements, divided into test cases, are performed:

- Test case 1: Radiation characteristics
- Test case 2: Signal strength
- Test case 3: Activity of the transmitter
- Test case 4: Detection of Secondary Vehicle Identifier in stand still mode of the vehicle
- Test case 5: Detection of Secondary Vehicle Identifier in driving mode of the vehicle

B. *Test Vehicles*

The following vehicles were investigated in the following measurements:

- Skoda Octavia 3: Only used for Bluetooth measurements
- VW Passat B8: Only used for Bluetooth measurements
- Opel Astra 2016 incl. OnStar: Only used for Wi-Fi measurements
- Opel Insignia Innovation 2016 incl. OnStar: Only used for Wi-Fi measurements

C. *Test Equipment*

1) Bluetooth Test Equipment:

- Notebook
  - ThinkPad X201 with Kali Linux (64 Bit, version 2016.2), BTScanner version 2.0, and Kismet version 2016-07-R1
  - Ubertooth One (firmware git-579f25) with Ubertooth-Specan-Ui, and Ubertooth-Rx version 201-10-R1 [13]
  - Standard antenna, LogPer Antenna and directional antenna WIFI-LINK WAVEGUIDE Antenna PN: WCA-2450-12, 2,4-2,5 GHz, 12 dBi
- Smart phone
  - Samsung Galaxy S6, Android 6.0.1, Bluetooth-Scanner app version 1.1.3 (from google playstore)

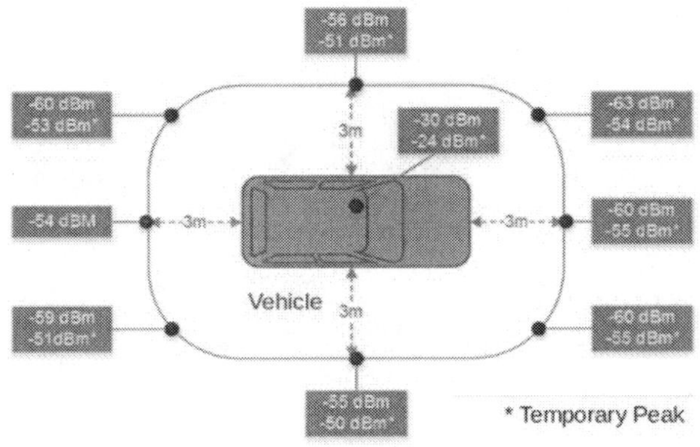

Figure 1. Radiation characteristic of the Octavia Bluetooth device

TABLE II. SIGNAL STRENGTH OF THE OCTAVIA BLUETOOTH DEVICE

Distance	Standard Antenna	LogPer Antenna	Directional Antenna
3 m	-50 dBm	-56 dBm	-47 dBm
6 m	53 dBm	-60 dBm	-51 dBm
9 m	-63 dBm	-63 dBm	-54 dBm
12 m	-67 dBm	-65 dBm	-56 dBm
15 m	-71 dBm	-68 dBm	-60 dBm
18 m	-75 dBm	-69 dBm	-63 dBm
21 m	-78 dBm	-72 dBm	-65 dBm
30 m		-75 dBm	-68 dBm

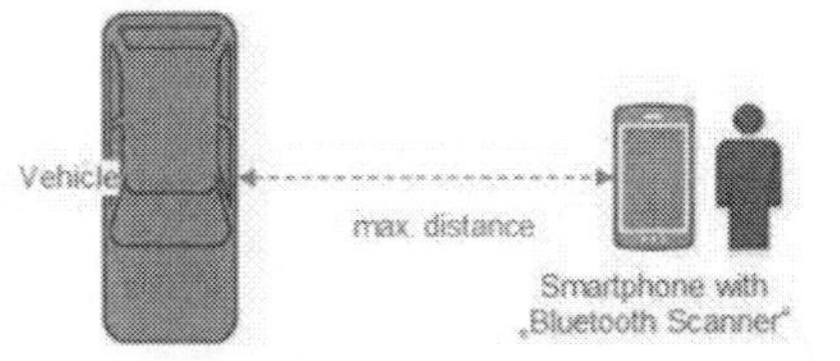

Figure 2. Test arrangement for the detection of Secondary Vehicle Identifier in stand still mode

2) Wi-Fi Test Equipment:

- Notebook
  - Notebook Lenovo ThinkPad T400, Ubuntu 16.04 LTS and LinSSID version 2.7
  - USB-Wi-Fi-device: TP-Link TL-WN722N with standard antenna and directional antenna WIFI-LINK WAVEGUIDE Antenna PN: WCA-2450-12, 2,4-2,5 GHz, 12 dBi
- Smart Phone
  - Huawei P8 lite 2017, Wifi-Analyzer App (from google playstore)
  - Samsung S7, Wifi-Analyzer App (from google playstore)

VI. MEASUREMENTS AND RESULTS

In this section the test results of the performed tests are desribed.

A. Bluetooth Measurements for the Octavia (and partly Passat)

1) Test Case 1: As test equipment a Lenovo ThinkPad X201, with Ubertooth One, Ubertooth-Specan-Ui and standard antenna is used. Measurements are performed at one position inside and 8 positions outside the vehicle. The positions and results are plotted in Figure 1. As we expected, the highest signal strength of -30 dBm has been detected inside the vehicle. But also outside the vehicle, a strong signal strengh has been measured.

2) Test Case 2: As test equipment a Lenovo ThinkPad X201, with Ubertooth One, Ubertooth-Specan-Ui and different antennas is used: Standard antenna, LogPer antenna and directional antenna WIFI-LINK WAVEGUIDE. The test results are presented in Table II. With all antennas the Bluetooth signal can always be detected, within a distance of 21 m.

3) Test Case 3: The Bluetooth module of the head-unit starts with scanning of Bluetooth devices which were already paired in the past and are registered in the pairing list of the head-unit after starting the ignition. Scanning is switched off after the deactivation of the ignition and removal of the key.

4) Test Case 4: First, as test equipment a Samsung Galaxy S6 with the Bluetooth scanner app is utilised. Figure 2 presents the test setting. The following information about the Bluetooth device of the head-unit can be captured with the mentioned test equipment:

```
Skoda_TF
00:17:CA:D9:6B:77 (−65 dBm)
AUDIO_VIDEO_HANDSFREE
Scan Cycle 199 (20.11.16 15:01)
```

SSID "Skoda_TF", BSSID "00:17:CA:D9:6B:77", the service "AUDIO_VIDEO_HANDSFREE" and the "Scan Cycle 199 (20.11.16 15:01)" with date were captured. These information are readable up to a distance of 24 m (signal strength at this distance: -83 dBm) (it has to be mentioned that the owner of the Skoda Octavia has already altered its SSID. "Skoda_TF" is not the factory setting).

The following information are captured from the Bluetooth device of the head-unit of the Passat up to a distance of 12 m (signal strength at this distance: -84 dBm):

```
VW BT 2058
A8:54:B2:FE:30:35 (−79 dBm)
AUDIO_VIDEO_HIFI_AUDIO
Scan Cycle 25 (02.11.16 13:15)
```

From a privacy perspective it is remarkable, that the name of the automaker is part of the SSID and that the number part "2058" of the SSID is chosen from the VIN of the Passat.

Next, as test equipment Lenovo ThinkPad X201, Ubertooth One with Ubertooth-RX is used to perform the same test case. Subsequent information can be captured if the test equipment is switched on and a Samsung Galaxy S6 will be connected to the Octavia head-unit:

```
systime=1479652524 ch=39 LAP=d96b77 err=0
clkn=100728 clk_offset=1540 s=−35 n=−55 ...
systime=1479652571 ch=39 LAP=68dae3 err=0
clkn=250437 clk_offset=5596 s=−21 n=−55 ...
systime=1479652571 ch=39 LAP=68dae3 err=0
```

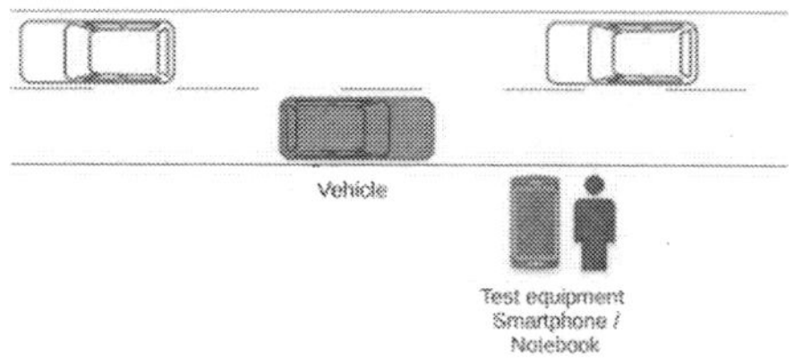

Figure 3. Test arrangement for the detection of secondary vehicle identifier in driving mode

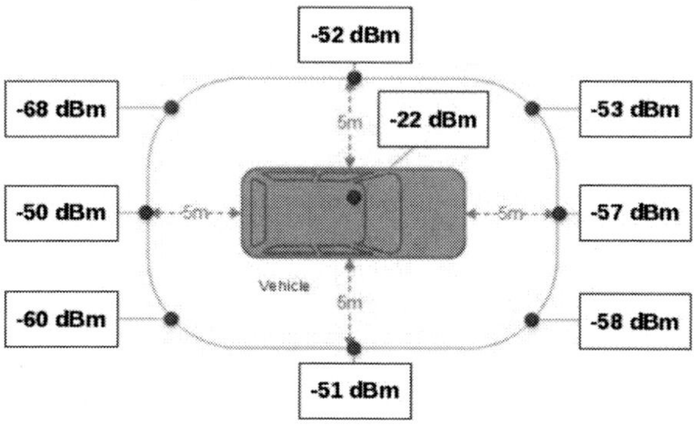

Figure 4. Radiation characteristic of the Opel Insignia Wi-Fi device

```
clkn=251217 clk_offset=5613 s=-16 n=-55 ...
```

This information can be captured up to 18 m with the standard antenna and up to 42 with the directional antenna.

5) Test Case 5: With the test equipment Samsung Galaxy S6 with the Bluetooth scanner app, subsequent information can be captured up to a speed of 30 km/h. Figure 3 shows the test case.

```
Skoda_TF
00:17:CA:D9:6B:77 (-65 dBm)
AUDIO_VIDEO_HANDSFREE
Scan Cycle 199 (20.11.16 15:01)
```

B. Wi-Fi Measurements for the Opel Insignia (partly Opel Astra)

1) Test Case 1: As test equipment a Lenovo ThinkPad T400, TP-Link TL-WN722N with standard antenna, and LinSSID is used. The signal strength of the Wi-Fi access point (Wi-Fi-AP) has been measured at 8 fix point outside and at 1 point inside the vehicle. The positions are equal to the Bluetooth test case. But in contrast to the Bluetooth measurement, the distance between the vehicle and the measurement tool is 5 m. The results for the Opel Insignia are plotted in Figure 4. As we expected, the highest signal strength of -22 dBm has been detected inside the vehicle. But also outside the vehicle, a strong signal strengh has been measured.

2) Test Case 2: As test equipment a Lenovo ThinkPad T400, TP-Link TL-WN722N with standard antenna, and LinSSID on the one side and Samsung S7, and Wifi-Analyzer on the other side are used. With the TP-Link TL-WN722 and

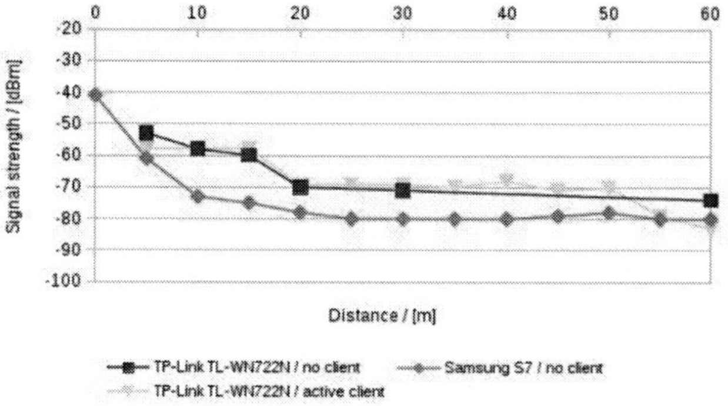

Figure 5. Radiation characteristic of the Opel Insignia Wi-Fi device

the Samsung S7 the signal strenght are measured in ascending distances to the vehicle, in direction to the right front door. The results are plotted in Figure 5. Only little differences in signal strength can be detected between an active connection and a non connection of a client to the Wi-Fi-AP of the Opel Insignia. The measurement sensitivity of the smart phone is about 10 dBm lower for distances greater 10 m in contrast to the measurements with the TP-Link. With both measurement devices the signal of the Wi-Fi-AP can always be detected, within a distance of 60 m.

3) Test Case 3: As test equipment a Lenovo ThinkPad T400, TP-Link TL-WN722N with standard antenna, and LinSSID is used.

General Motor and Opel provide vehicle online connectivity based on the OnStar service. Only if the OnStar service is enabled the Wi-Fi-AP of the Opel Insignia can be switched on. The Wi-Fi transmitter is activated, when the ignition is started and deactivated when the key is removed from the ignition lock. An enabling or disabling of the Wi-Fi-AP is not possible by the driver, with the usage of the configuration menu implemented in the vehicle (disabling possible with an appropriate smartphone app).

4) Test Case 4: As test equipment a Lenovo ThinkPad T400, TP-Link TL-WN722N with standard antenna, and LinSSID on the one side and Samsung S7, and Wifi-Analyzer on the other side are used. Figure 2 presents the test setting. In stand still mode the following Secondary Vehicle Identifier and additional information has been measured for the Wi-Fi device of the Opel Insignia, for all distances up to 60m with both test equipments.

```
SSID: WiFi Hotspot 1760
BSSID: C4:49:BB:21:91:DE
Frequency: 2437 MHz; 2448-2426 = 22 MHz
Channel: 6
Misc.: WPA2-PSK-CCMP+TKIP, ESS,
       MITSUMI ELECTRIC Co.,LTD
```

Next, we determine the maximum detection distance for the Secondary Vehicle Identifier. As test equipment a HP notebook, TP-Link TL-WN722N with standard antenna, and a LinSSID on the one side and a Huawei P8 lite 2017 with a Wifi-Analyzer on the other side are used. The results are presented in Table III for the Wi-Fi device of the Opel Astra.

TABLE III. SIGNAL STRENGTH OF THE ASTRA WI-FI DEVICE IN STAND STILL MODE

Distance	Signal strength Huawei P8 lite 2017	Signal strength TP-Link TL-WN722N
216 m	-82 dBm	-81 dBm
424 m	no signal	-91 dBm

TABLE IV. SIGNAL STRENGTH OF THE ASTRA WI-FI DEVICE IN DRIVING MODE

Speed	Maximum signal strength Huawei P8 lite 2017	Maximum signal strength TP-Link TL-WN722N
50 km/h	-60 dBm	-55 dBm
100 km/h	-71 dBm	-50 dBm

If a signal has been detected, then the SSID and the BSSID can always be extracted. The smart phone detected a signal up to 216 m, the USB-Wi-Fi-device up to 424 m.

5) Test Case 5: As test equipment a HP notebook, TP-Link TL-WN722N with standard antenna, and a LinSSID on the one side and a Huawei P8 lite 2017 with Wifi-Analyzer app on the other side are used. Notebook with USB - Wi-Fi device and smart phones operate 1 m above the floor beside the roadway. Figure 3 shows the principle test case. The results for the Wi-Fi device of the Opel Astra are presented in Table IV. The maximum signal strenght has been detected by the USB-Wi-Fi-device. The measured signal strengths with the TP-Link for 50 and 100 km/h are surprising. We assume that this issue is caused by the moving vehicle and the sample rate of the measurement devices of about 1 Hz (vehicle moves 13,9 m/s by 50 km/h and 27,8 m/s by 100 km/h).

VII. CONCLUSION

As shown in Section VI, it is technically very easy to capture Secondary Vehicle Identifier based on wireless interfaces of vehicles, especially Bluetooth and Wi-Fi (even with low cost equipment shown in this paper). Although, this interfaces are designed to connect devices of occupants, vehicle identifier can be detected far away from the vehicle (Wi-Fi 424 m with a TP-Link device) and high vehicle speed of up to 100 km/h. This enables the misuse of vehicle identifier for the tracking of vehicles.

In the context of the upcoming V2V communication our results are worrying concerning privacy of vehicles and drivers. The V2V communication is a short range communication technology with a communication range of about 800 m in open space. In future, each vehicle periodically broadcasts Cooperative Awareness Messages (CAM) with a packet generation rate of 1 up to 10 Hz. A CAM contains a lot of data about the sending vehicle: current geographic position, speed, driving direction, etc., at a specific time. One privacy requirement is that a receiver can not link a CAM to a specific vehicle. Now, Secondary Vehicle Identifier can be misused to link captured CAM messages to a specific vehicle [14].

Besides the investigation of wireless Secondary Vehicle Identifiers, we noticed, that the security configuration of the Wi-Fi-APs in the examined vehicles should be improved. For example, neither MAC filtering, invisibility of the SSID identifier nor adjustable signal strengh, etc., can be set based on the Wi-Fi-AP configuration menues. A first improvement is to implement the security features suggested for WiFi-APs in generell. A comprehensive source for network security is the BSI IT-Grundschutz Catalogue [15].

ACKNOWLEDGEMENT

The authors would like to thank our colleague Thomas Strubbe for detailed discussions. Also thanks to the anonymous reviewers for the valuable comments.

REFERENCES

[1] Markus Ullmann, Thomas Strubbe, and Christian Wieschebrink, "Technical Limitations, and Privacy Shortcomings of the Vehicle-to-Vehicle Communication," in Proceedings VEHICULAR 2016: The Fifth International Conference on Advances in Vehicular Systems, Technologies and Applications. IARIA, 2016, pp. 15–20.

[2] H. Lee, D. Kim, D. Kim, and S. Y. Bang, "Real-time automatic vehicle management system using vehicle tracking and car plate number identification," in Multimedia and Expo, 2003. ICME'03. Proceedings. 2003 International Conference on, vol. 2. IEEE, 2003, pp. II–353.

[3] R. Bruno and F. Delmastro, "Design and Analysis of a Bluetooth-Based Indoor Localization System," 2003, pp. 711–725.

[4] R. Zhou, "Wireless Indoor Tracking System (WITS)," Aktuelle Trends in der Softwareforschung, Tagungsband zum IT Software-Forschungstag. Dpunkt Verlag Heidelberg, Germany, 2006, pp. 163–177.

[5] C. Carpenter, M. Fowler, and T. Adler, "Generating Route-Specific Origin-Destination Tables Using Bluetooth Technology," Transportation Research Record: Journal of the Transportation Research Board, no. 2308, 2012, pp. 96–102.

[6] M. Emery and M. K. Denko, "IEEE 802.11 WLAN Based Real-Time Location Tracking in Indoor and Outdoor Environments," in Electrical and Computer Engineering, 2007. CCECE 2007. Canadian Conference on. IEEE, 2007, pp. 1062–1065.

[7] R. M. Ishtiaq Roufa, H. Mustafaa, S. O. Travis Taylora, W. Xua, M. Gruteserb, W. Trappeb, and I. Seskarb, "Security and privacy vulnerabilities of in-car wireless networks: A tire pressure monitoring system case study," in 19th USENIX Security Symposium, Washington DC, 2010, pp. 11–13.

[8] S. Astapov and A. Riid, "A Multistage Procedure of Mobile Vehicle Acoustic Identification for Single-Sensor Embedded Device," International Journal of Electronics and Telecommunications, vol. 59, no. 2, 2013, pp. 151–160.

[9] B. Danev, D. Zanetti, and S. Capkun, "On Physical-Layer Identification of Wireless Devices," ACM Computing Surveys (CSUR), vol. 45, no. 1, 2012, p. 6.

[10] ETSI, "ETSI TR 102 893 V1.1.1: Intelligent Transport Systems (ITS); Security; Threat, Vulnerability and Risk Analysis (TVRA); Technical Report," 2010, http://www.etsi.org/, Access Date: June 02, 2017.

[11] ——, "ETSI EN 302 665 V1.1.1: Intelligent Transport Systems (ITS) - Communications Architecture," 2010, http://www.etsi.org/, Access Date: June 02, 2017.

[12] Bluetooth Special Interest Group, "Specifications," 2017, https://www.bluetooth.com/specifications, access date: March 24, 2017.

[13] Ubertooth Developer, "Ubertooth Bluetooth Sniffer," 2017, https://github.com/greatscottgadgets/ubertooth/, access date: March 24, 2017.

[14] Markus Ullmann, and Thomas Strubbe, and Christian Wieschebrink, "Misuse Capabilities of the V2V Communication to Harm the Privacy of Vehicles and Drivers," in International Journal On Advances in Networks and Services, vol 10 no 12. IARIA, 2017.

[15] Federal Office for Information Security, "IT-Grundschutz Catalogues," 2013, https://www.bsi.bund.de/EN/Topics/ITGrundschutz/ITGrundschutz Catalogues/itgrundschutzcatalogues_node.html, access date: March 30, 2017.

Security Hardening with Plausibility Checks for Automotive ECUs

Jürgen Dürrwang, Marcel Rumez, Johannes Braun and Reiner Kriesten

Institute of Energy Efficient Mobility
University of Applied Sciences Karlsruhe
Germany, International University Campus 3, 76646 Bruchsal
Email: {duju0001, ruma0003, brjo1015, krre0001}@hs-karlsruhe.de

***Abstract*—The automotive industry relies increasingly on computer technology in their cars, which malicious attackers can exploit. Latest published attacks have further shown an increased attack surface by adding wireless interfaces to vehicle on-board systems. Most of these attacks are based on spoofing or sending tampered bus messages, which we were able to reproduce over the last years as well. We found additional vulnerabilities with the same attack vector in cars of international Original Equipment Manufacturers (OEMs). The discovered vulnerabilities can be dangerous to life while the driver doesn't have any possibilities to prevent them. Based on this knowledge we developed an approach to prevent such attacks on Electronic Control Unit (ECU)-level. In this publication, we introduce a new type of countermeasure to reduce the attack surface of vehicles with less or no overhead. Therefore, we concentrate on plausibility checks in a new way, by employing hard-wired signals to determine the operational state of the car. As a result, we are hardening the security against attacks on legitimate functions.**

Keywords–Automotive Safety and Security; Vehicular Attacks; Plausibility Checks.

I. INTRODUCTION

Modern automobiles consist of more than 50 ECUs, which contain and implement a total of up to 100 million code lines to control safety-critical functionality. This fact, combined with the close interconnectivity of automotive ECUs, opens up new possibilities to attack these systems which impair the safe operation of the vehicle. The feasibility of such attacks has been investigated and already demonstrated by several groups of researchers [1] [2]. Additionally, attacks via access to the internal vehicle network, that can cause life-threatening injuries have also been demonstrated in the past [3] [4].

Furthermore, car manufacturers tend to equip their cars with more entertainment and comfort features using wireless connectivity. One example is the detection of traffic obstructions by using Car-2-X communication to process traffic or general environmental information provided by an ad-hoc network. In the same way, providers of car-sharing, car-rental and other fleet based services use cellular networks for the communication with their backbone. Additionally, manufacturers implement the ability to execute software updates outside of car workshops, in order to fix problems within a short time [5]. These interfaces potentially provide means to remotely exploit vulnerabilities, obtain access to the in-vehicle network and control critical systems from a distance [6] [7].

Especially, with the remote exploitation of the Jeep Cherokee [6], Miller and Valasek showed that physical access through an On-Board Diagnostics (OBD)-Connector is not mandatory any more. One year after the remote exploitation of the Jeep they provided an update on what is possible in car hacking. This time the experts didn't use a remote connection for their attacks, but a direct connection to the internal car network via the OBD-connector. The fundamental approach was to stop an ECU which is connected to the Controller Area Network (CAN) in order to send spoofed messages to another in-vehicular subscriber. As a result, they were able to execute different functions, e.g., deceleration of the vehicle or activating the parking assistant in an inappropriate driving condition. To prevent such misuses, ECUs typically use plausibility checks to validate the requested function with the state of vehicle. For this purpose ECUs use bus messages to derive the current state of the vehicle. Unfortunately, these messages are typically not protected from malicious modifications.

Additionally, our actual research has discovered a weakness in a safety critical component due to the fact, that this component provides diagnostic functions for a special use case. Unfortunately, these functions are available during the regular operation of the vehicle, potentially leading to life-threatening injuries. The discovered weakness is based on a requirement, suggesting a weak algorithm to ensure authentication. Moreover, this requirement is proposed by a standard. Thus, we consider it as reasonable, that this weakness scales over several manufacturers.

To prevent such issues, authenticity and integrity of bus messages has to be ensured and therefore cryptographic methods can be applied. A typical approach for this is the application of a Keyed-Hash Message Authentication Code (HMAC) on salted messages. This type of cryptographic measure ensures the desired protection goals, with an acceptable need of computational performance, which is a fundamental constraint in the automotive domain. Nevertheless, there are existing drawbacks when using HMACs. In particular, the increasing bus load when attaching an HMAC on each message. Furthermore it requires an extensive key management. Accordingly to the constraints in the automotive domain like restricted bandwidth and power, a trade-off between protection level and required resources is necessary. Unfortunately, this often leads to a non-implementation of necessary security measures. In this paper, we propose an approach of using local ECU signals, in addition to the information which the ECU receives from bus systems, to perform plausibility checks. In detail, the contributions of this paper are the following:

Problem: Spoofing and tampering of bus messages in vehicular networks can lead to safety critical situations. To prevent these situations, message authenticity and integrity have to be ensured. Therefore, cryptographic measures can be used, but they are often not applicable due to the fundamental constraints in the automotive domain. **Solution:** Apply plausibility checks with local ECU signals to verify data integrity without cryptography. Our **Contribution:** A novel approach for plausibility checks with local or directly measured signals for hardening security in the automotive domain. Moreover, the approach to secure safety-critical functions with plausibility checks hardens security with minimal integration effort in the typical automotive engineering process.

The paper is structured as follows: Section II summarizes the related work in the area of automotive security measures, followed by our approach in Section III, which is divided in methodology and its applicability. Furthermore, we propose a way to locate suitable signal sources inside vehicles that are necessary for our approach. This is followed by an application example that should be able to prevent the published exploitation of a passenger vehicle. In Section IV we give a short summary of our work and present an outlook on how our approach could be combined with other security measures in Section IV.

II. Related Work

Automotive manufacturers, suppliers and other organizations have already recognized the necessity for security mechanisms in the automotive domain. For this reason, a cyber security alliance was founded in the USA. The major objective of the Automotive Information Sharing and Analysis Center (AUTO-ISAC) [8] is to enhance cyber security awareness and the coordination for the automotive domain. Moreover, the alliance is providing best practices for organizational and technical security issues to support the developing process of their members. An additional effort was initialized by the Society of Automotive Engineers (SAE) with the J3061 guidebook [9], summarizing recommended security practices that can be applied in the automotive domain. Unfortunately, the guidebook gives no concrete reference implementations for possible measures.

A more comprehensive approach for security in cars is presented by Gerlach et. al. [10]. They propose a multi-layer security architecture for vehicular communication which implements different measures. In particular, they propose digital signatures with certificates as methods for providing authentication, integrity, and non-repudiation of the received messages. Due to the underlying asymmetric cryptography, high-performance ECUs or ECUs with additional Hardware Security Modules (HSMs) are needed. They further consider an application of cross-layer plausibility checks [10] as meaningful. Therefore, they establish a single instance in the vehicle which collects information from any existent source in the car. The instance is called plausibility checking module and creates its own independent view of the current vehicle state. If deviations from normal operation are detected, the instance reacts by triggering a warning. Unfortunately, the proposed instance is not implemented in each ECU, hence triggered counteractions or warnings have to be transferred over the unsecured bus again.

An additional approach is presented by Dhurandher and his researchers [11]. They propose an application of reputation and plausibility checks for Vehicular Ad Hoc Networks (VANETs). In particular, their proposed algorithm is able to detect and isolate malicious nodes by the use of sensors. Although they present an efficient and effective algorithm, the approach is designed for wireless nodes and their unique characteristics. Unfortunately, a concept for adaptation to in-vehicle networks is not given.

III. Approach

We consider an application of plausibility checks as additional protection mechanism as meaningful, if the relevant functions are able to change the physical state of the vehicle. This is partly explained by the fact that for these type of functions sensor values are already exist. As a result, our approach is applicable for a great set of functions and in particular for almost all safety-related functions. To decide if a function can be protected by our approach, some requirements have to be met. We define these requirements in the following and we further present an application example. Therefore, we divide our approach into two parts: (1) required steps to validate if the selected function can be protected by a plausibility check (see Figure 1) and (2) a reference implementation for plausibility checks with local ECU signals. Finally, we give an application example which is explained in Section III-C).

A. Applicability of Plausibility Checks

To validate if plausibility checks are applicable, a few requirements have to be checked beforehand. For this purpose, we define and highlight them as selection steps in Figure 1.

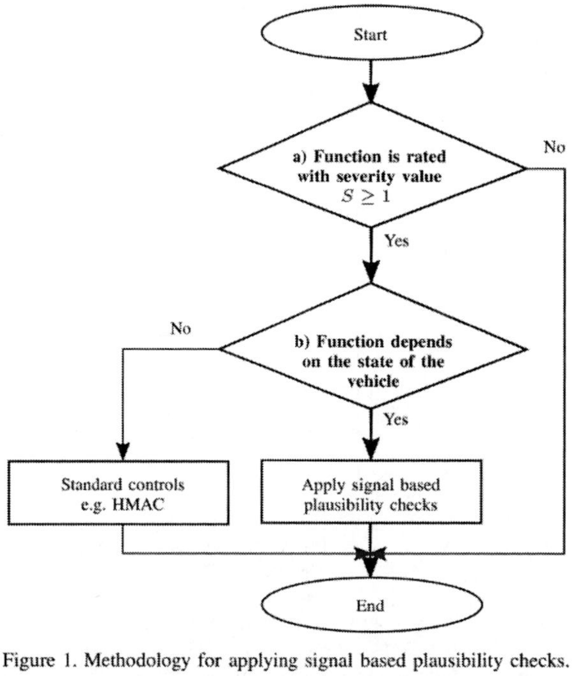

Figure 1. Methodology for applying signal based plausibility checks.

Figure 1 shows the required steps to identify functions that are applicable for plausibility checks. Before we can validate *Step a)*, a hazard and risk analysis has to be performed. This is a demand of the functional safety standard ISO 26262 [12]. The aim of the analysis is to identify potential hazards of a function. Furthermore, for each hazard a so-called Automotive Safety Integrity Level (ASIL) based on three values is calculated. One of these values is defined as severity, describing the possible impact of the malfunction related to the selected function. Thus, we consider a selection of functions able to cause hazards with a severity value greater or equal to S1 as meaningful. In particular, a severity value of $S \geq 1$ implies injuries of vehicle occupants [12] and must be prevented. If the function is rated with $S \geq 1$, the next step is to check, if the selected function has dependencies on the vehicle state (moving or standing still, etc.) as shown by *Step b)* in Figure 1. If plausibility checks are not applicable, but the function is rated with $S \geq 1$, we deem an application of standard security controls to be mandatory.

B. Plausibility Checks with Local ECU Signals

To guarantee that signals used for plausibility checks can not be maliciously modified or sent, we have to implement protection mechanisms. In particular, we have to ensure the authenticity and integrity of the used signals. Therefore, we could apply the already mentioned cryptographic methods with all their drawbacks. Instead, we chose another way to ensure the protection of the information assets without the afore mentioned drawbacks. To explain the approach we take a closer look into automotive architectures like the one presented in Figure 2.

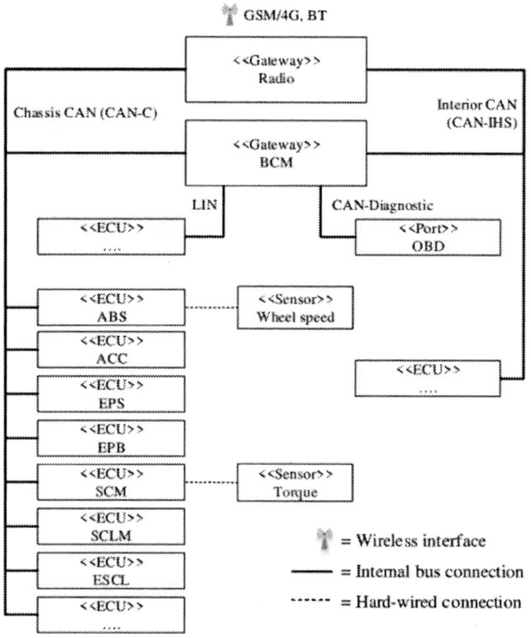

Figure 2. Part of the electrical architecture of a Jeep Cherokee 2014, based on the work of Valasek et. al. [6]. As diagram notation we use the UML4PF profile extension [13].

Figure 2 represents a part of the E/E architecture of a Jeep Cherokee 2014, which was the attack target of the researchers [6] [14] mentioned in the beginning. The architecture shows different ECUs and gateways interconnected by three CAN-Bus systems (CAN-C, CAN-IHS, CAN-Diagnostic), as well as one LIN-Bus. Furthermore, each wheel has a sensor measuring the wheel speed which is hard-wired to the Antilock Braking System (ABS), respectively the Electronic Stability Control (ESC). This information can be used to derive ECU local signals for plausibility checks without the need for cryptographic algorithms. In particular, these sensor values can indirectly describe the state of the vehicle. With the wheel speed sensor shown in Figure 2, we can derive whether the vehicle is moving or not. If the vehicle is on halt, all sensor values of the wheels have to be zero or vary significantly due to a spinning wheel. This hard-wired sensor type is only an example. Additionally, we can combine two or more sensor values to derive more precise information about the state of the vehicle. The important point in our approach is that an ECU with hard-wired sensors can operate as a guardian against spoofed or tampered signals. In general, it is important that a safety critical function can be additionally protected by one or more hard-wired sensor values. By adding this requirement, an attacker would no longer be able to spoof sensor values over bus messages, while ECUs can verify the plausibility of the received values. We are aware that hard-wired sensors can increase cabling efforts, if an ECU normally doesn't have access to any hard-wired sensors. However, the addition of security techniques is often tied to increased costs.

To be precise, authenticity and integrity are only ensured, if the attacker is not capable of getting access to the sensors themselves, requiring him to be in the vicinity of the vehicle. We assume that the possibility of an attacker accessing sensors is unlikely in comparison to his ability to send spoofed messages via CAN [6]. This is reasonable due to the fact that an attacker has to overcome several physical barriers, e.g. opening the hood, ECU housing or removing the wire insulation.

C. Application Example

As an example, we want to discuss the latest Jeep [14] hack, as well as the attack on the steering system which have been done. Generally, the vulnerabilities in diagnostic mode, which the researchers used for disabling the Jeep's brakes among other things, are only working if the car is in reverse and slower than 5 mph. How can we make sure, that the values received for plausibility checks are valid and not tampered with? We want to answer this question by the following examples, based on the already mentioned vulnerabilities and how our approach would prevent these hacks in the future.

In the first example, the researchers set the real ECU in a service mode causing it to stop sending messages on the bus. This step enables them to send their own messages in the name of the jammed ECU. Electric Power Steering (EPS), which can be integrated in modern vehicles, e.g., the hacked Jeep series, requires various input parameters for calculating the electric steering support. One of these control values is the velocity of the vehicle. Depending on the current speed and other parameters, the Steering Control Module (SCM) calculates the steering torque. Basically, the steering torque support is decreasing by the SCM, when the velocity is increasing. Applied to the example of the Jeep hack, we want to show the determination of the steering torque threshold, which was one of the conditions the Jeep had to meet, in order to execute the steering angle change. A request for a high torque support in vehicle speeds of 30 mph or higher is not legitimate. However, we have to ensure that the integrity of the velocity value is given, for example by a hard-wired connection of the wheel speed sensors to the SCM. For instance, by implementing our approach, we deem the execution of the function as done in the hack would have been refused during the plausibility check.

Another attack presented by Valasek and Miller [14] was the application of the car's brakes. The exploited function is normally used to activate the electronic parking brake for emergency braking by pressing the parking switch for a longer amount of time. Thereupon the pump for the ABS and ESC system gets activated and provides the necessary pressure to engage the brakes of the car. In this case, our approach is not applicable because of the missing hard-wired signals. In particular, an implemented plausibility check would not be possible, because of the lack of hard-wired signals. Therefore it can not be differentiated between unintended or intended emergency braking, because we have only the information from the bus. In a case like this, where no hard-wired signal sources available we propose to check the feasibility of adding a hard-wired connection.

Our own attempts have shown, that the related safety relevant ECU mentioned in the introduction has already connected hard-wired signals. However, the existent checks do not analyse the use-case correctly. Thus, it would have been possible to increase the security level simply by using enhanced software prompts, e.g., logical *and/or* conjunctions.

IV. Conclusion

In this publication, we proposed a new way to implement plausibility checks for automotive ECUs. The approach is capable to ensure that signals used for plausibility checks are resilient against replay and tampering. Furthermore, the approach uses already available information, like sensor signals, to verify function requests with the actual state of the vehicle. Due to the fact, that no cryptography is needed and existent information is reused, our approach requires less performance and costs, e.g., no HSM chips, as well as no additional busload than other security measures. Additionally, we showed an example implementation of our approach, which is able to prevent a known attack. We used hard-wired sensor signals like wheel speed sensors of the ABS to ensure the integrity of the velocity signal. Furthermore, using the electric power steering ECU example, we have shown how a function is able to perform a plausibility check. This is done by a function request during runtime using characteristic values.

After doing our own research we can confirm that replay attacks can be performed with minimal effort, if bus systems like CAN are used. In combination with our findings based on a safety critical function in an ECU, which is rated with a severity value of 3, we recommend that such functions should only be executable by bus messages if they validate the plausibility of the request. Therefore, our approach recommends using at least two values received from different sources. In the best case scenario, one source is a hard-wired connection. Finally, messages for ASIL D functions should not be routed over gateways, unless there is no other way. This will prevent relaying malicious messages between different domains.

V. Future Work

The mentioned vulnerabilities show us the necessity of additional safeguards for future vehicles. This creates new challenges for the whole automotive domain, in addition to the rising amount of interconnected services. Due to this fact, we are working on a following step including distributed firewalls techniques to enhance the security level for diagnostic services and functions which can change the physical state of the car, by using Stateful Packet Inspection (SPI), as well as Deep Packet Inspection (DPI). We want to use the approach to distinguish between different sets of implemented policies depending on the requested use case, e.g., starting a diagnostic session. For validating the specific use case the internal firewalls have to check different hard-wired sensors like wheel speed or seat occupancy. Thus, the firewalls have to ensure that requested services are matching to the associated use cases based on the vehicle state. Additionally, through the use of SPI, we assume that the firewall would be able to detect anomalies, like the attack example of jamming an ECU by setting it in boot ROM mode in order to spoof its control functions influencing the vehicle movement. We are convinced that plausibility checks with local signals are hardening the security of a vehicle. Finally, we're going to publish detailed information regarding the weakness in the safety relevant ECU in the near future and contribute an enhanced approach fixing this type of issue. Additionally, the mentioned firewall techniques shall also be evaluated and flow into an additional publication.

Acknowledgment

This work has been developed in the project SAFE ME ASAP (reference number: 03FH011IX5) that is partly funded by the German ministry of education and research (BMBF) within the research programme ICT 2020.

References

[1] S. Checkoway, D. McCoy, B. Kantor, D. Anderson, H. Shacham, S. Savage, K. Koscher, A. Czeskis, F. Roesner, T. Kohno et al., "Comprehensive experimental analyses of automotive attack surfaces," 2011.

[2] C. Miller and C. Valasek, "A survey of remote automotive attack surfaces," black hat USA, 2014, last checked on 09.05.2017. [Online]. Available: https://sm.asisonline.org/ASIS%20SM%20Documents/remote%20attack%20surfaces.pdf

[3] K. Koscher, A. Czeskis, F. Roesner, S. Patel, T. Kohno, S. Checkoway, D. McCoy, B. Kantor, D. Anderson, H. Shacham et al., "Experimental security analysis of a modern automobile," Symposium on Security and Privacy, 2010.

[4] C. Miller and C. Valasek, "Adventures in automotive networks and control units," DEF CON, vol. 21, 2013, pp. 260–264.

[5] M. S. Idrees, H. Schweppe, Y. Roudier, M. Wolf, D. Scheuermann, and O. Henninger, "Secure automotive on-board protocols: A case of over-the-air firmware updates," Nets4Cars/Nets4Trains 2011, 2011, pp. 224–238.

[6] C. Valasek and C. Miller, "Remote exploitation of an unaltered passenger vehicle," last checked on 09.05.2017. [Online]. Available: http://illmatics.com/Remote%20Car%20Hacking.pdf

[7] A. Greenberg, "Tesla Responds to Chinese Hack With a Major Security Upgrade," last checked on 23.03.2017. [Online]. Available: https://www.wired.com/2016/09/tesla-responds-chinese-hack-major-security-upgrade/

[8] AUTO-ISAC, "Automotive information sharing and analysis center," https://www.automotiveisac.com/index.php, last checked on 09.05.2017.

[9] SAE, "Cybersecurity Guidebook for Cyber-Physical Vehicle Systems," 01.2016, last checked on 12.04.2016. [Online]. Available: http://standards.sae.org/wip/j3061/

[10] M. Gerlach, A. Festag, T. Leinmuller, G. Goldacker, and C. Harsch, "Security architecture for vehicular communication," in Workshop on Intelligent Transportation, 2007.

[11] S. K. Dhurandher, M. S. Obaidat, A. Jaiswal, A. Tiwari, and A. Tyagi, "Vehicular Security Through Reputation and Plausibility Checks," Systems Journal, IEEE, vol. 8, no. 2, 2014, pp. 384–394.

[12] ISO, "ISO 26262 – Road Vehicles – Functional Safety," 2011.

[13] D. Hatebur and M. Heisel, "A uml profile for requirements analysis of dependable software," in International Conference on Computer Safety, Reliability, and Security. Springer, 2010, pp. 317–331.

[14] C. Valasek and C. Miller, "CAN Message Injection: OG Dynamite Edition," last checked on 05.04.2017. [Online]. Available: http://illmatics.com/can%20message%20injection.pdf

Evaluation of a General Purpose Communication Unit for Two-Wheelers

José Santa, Pedro J. Fernández and Antonio F. Skarmeta
Department of Information and Communication Engineering
Computer Science Faculty, University of Murcia
30100 Murcia, Spain
Email: {josesanta, pedroj, skarmeta}@um.es

***Abstract*—Vehicular communications are a need in future smart scenarios, since people can spend hours on a variety of transport means in daily activities. However, most of the current research deals with the integration of telematics in light vehicles, while two-wheelers (or equivalent) present especial conditions that reveal a need for new communication units that allow integrating bikes and mopeds (among others) into vehicular networks. In this paper, this gap is solved with a new embedded design provided with IEEE 802.11p and cellular communications, which can work as a mobile router and presents a simple but effective interface for warning services. A key point of the proposal is the bet for IPv6 as the base network protocol, which will be the essential in all-connected environments following the IoT paradigm. The communication unit has been tested in a real driving scenario, and one-hop results using the 802.11p channel reveal communication delays bellow 2 ms, packet delivery rates above 50% within a road stretch of 400 meters, and a maximum throughput of 4 Mbps. These performance values are adequate for a number of safety, infotainment and exploitation services.**

Keywords–Communication unit; Two-wheeler; Performance evaluation; IPv6.

I. Introduction

Cooperative Intelligent Transportation Systems (C-ITS) are demonstrating to improve safety and mobility of common vehicles, as it has been reported in results of European projects like Drive C2X [1] or FOTsis [2]. These systems stretch the range of isolated systems powered with sensors to better maintain contextual information about the surrounding traffic, thanks to wireless communication technologies. Vehicular networks have reached a standardization stage and vehicle to vehicle (V2V) and vehicle to infrastructure (V2I) communications are considered to save lives and improve travel experience in the near future. The USA government has stated its deployment strategy in this line [3], for instance. However, the work in this area focused on two-wheelers is still limited in the ITS context, even though European studies remarked the need for further research in the area of vulnerable road users (VRU) [4]. Bikes, mopeds and motorbikes have special constraints that make them a different vehicle group to be integrated in the Future Internet. Power source, mobility patterns, vehicle dimensions, communication range, positioning capability, human-machine interface (HMI) or electronics integration are some of these particular features. The potential of C-ITS in the two-wheel segment is clear, but a proper technological platform is needed to provide effective safety and mobility services.

For C-ITS to happen, a network that interconnects all the different elements of the road is necessary. In this field, different standardization bodies have defined its own reference communication architecture. IEEE have bet on Wireless Access in Vehicular Environments (WAVE), whose most relevant technology in the last years has been 802.11p, also called *Vehicular WiFi*, which provides the physical and access layers of the stack. ETSI and ISO propose an architecture with an exchangeable access technology, and propose different routing protocols. What is remarkable in these three proposals is the presence of the IPv6 protocol, although it has received a marginal interest and it is mainly considered by these organizations as a complementary network protocol mainly for infotainment applications. However, we understand IPv6 as an essential piece of a vehicular communication stack to integrate vehicular networks in the Future Internet.

The work described in this paper follows the objective of providing a communication unit adapted to the special needs of two-wheelers by also using IPv6 as a reference interconnection protocol. For this, the proposal raises from the synergy between current ITS standards (ISO/ETSI), Internet protocols (IETF) and IEEE technologies. The solution is an embedded communication unit for two-wheel vehicles integrating IEEE 802.11p and 3G wireless technologies in a small-factor computer, and running a communication middleware based on IPv6. Apart from the design, the work is especially focused on the prototype of the platform and the real communication performance tests.

The paper is organized as follows. Section II places this work in the research literature. Section III describes the design and development of the new communication unit for two-wheelers, while Section IV focuses on the IPv6 communication middleware integrated. Section V includes the experimental evaluation of the unit. Finally, Section VI concludes the paper and describes the next steps of our work.

II. State of the Art

A proper scientific knowledge supporting the application of C-ITS technologies in two-wheelers is limited in the literature. The authors in [5] have recently discussed the importance of communications for enhancing safety and efficiency of VRU and, particularly, the protection of two-wheelers. Several works in the literature already deal with road safety regarding pedestrians, which are the ones that barely cite two-wheelers in the area of vehicular communications. In [6], a review of systems to protect pedestrians reveals that until 2007 communications were rarely used to create cooperative systems, and vision, thermal, radar or laser sensors were used to avoid vehicle collisions with VRU. The proliferation of lower-delay cellular connections and the wide usage of WiFi contributed to the appearance of cooperative solutions around 2010. In [7], the base idea of exchanging localization data among pedestrian mobile phones and car on-board devices is presented. Here, a 3G link is mainly used, although it is also discussed the establishment of a WiFi connection between both terminals to reduce communication delay. A similar system is presented in [8] and [9], although they analyze in more detail the implications of the communication technology used. The authors in [10] review a communication technology in the 700 MHz for pedestrian to vehicle communication, which is far from current USA and EU trend of working in the microwave band. The solution presented in this paper focuses on the use of vehicular WiFi technology in the 5 GHz band for short-range vehicular communications, whose potential for communications between cars and VRU is discussed in [11], and cellular networks for ubiquitous access to Internet when 802.11p is not available.

A recent work dealing with the integration of telematics in bicycles can be found in [12]. It is a preliminary system concept in which ZigBee is used to improve travel efficiency through cooperative cruise control in bikes. Although the communication system is apart from current vehicular standards, it shows an interesting equipment embedded in the bike with a haptic interface. In [13], a prototype of connected motorbike uses a communication device installed in the boot with vehicular WiFi (5 GHz). This is a reference research in the motorbike segment, although further work is identified in a protocol stack lacking support with current trendy standards in the segment and not considering Internet connection. The work in [14][15] is focused on light two-wheel vehicles, presenting a safety system to warn cyclist about the presence of other vehicles through a visual interfaced embedded in the helmet. However, as a difference with the present contribution, it is based exclusively on a 3G/4G connection through the mobile phone for the case of the bike. The work in [16] presents a system that also uses a mobile phone for a similar purpose, although the novelty in this case is found in the way that regular WiFi is used. Beacon messages are used to directly embed information about a safety service, thus avoiding the association. A drawback of this proposal is its narrow application perspective and the need of using this especial communication mode of WiFi. A work including standardized vehicular communication protocols is described in [17], where motorbikes are equipped with a unit using an ETSI-compliant communication stack, while bikes use a especial bluetooth low-power device. Although this work share with the present contribution the idea of connecting two-wheelers to the vehicular network, it is focused on especial protocols for vehicular networks, and thus it sets IPv6 aside. The proposal in [18] does not integrate up-to-date technologies in the vehicular segment, but proposes the usage of IPv6 in the bike domain as a proof of concept.

As can be seen, existent contributions that especially tackle two-wheelers connectivity are generally based on technologies and protocols far from dealing with the interconnection of these transportation means with the rest of the Internet. We think that IPv6 is the key piece of smart environments (e.g., smart cities) and, for sure, of Future Internet architectures and, for this reason, we present a communication unit able to integrate two-wheelers in IPv6 networks with a hardware adapted to this operation scenario.

III. Hardware Platform

A new communication unit has been created for the case of two-wheel vehicles, including necessary communication technologies and providing a proper software host. The unit was

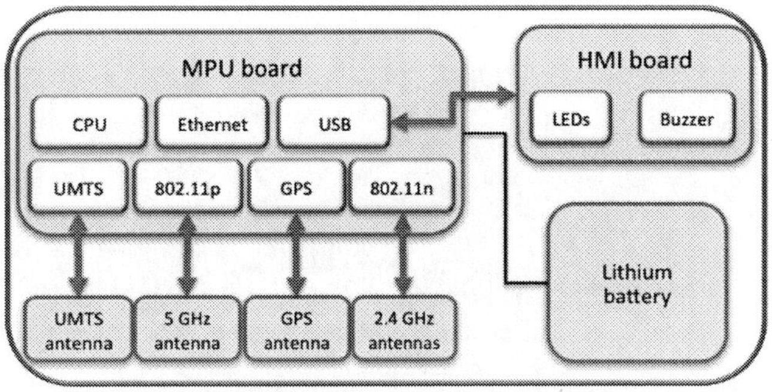

Figure 1. Hardware design of the communication unit

initially presented in [19], considering a basic communication middleware for a constrained safety application, and presenting an initial prototype. The platform has been now improved to support IPv6 network mobility to act as mobile router and the prototype has been assembled in its final form, being totally operational, as it is demonstrated in Section V.

The general architecture of the unit is depicted in Figure 1. The microprocessor unit (MPU) board contains the base platform, with the CPU, USB, and communication modules supporting Ethernet, UMTS (3G), vehicular WiFi, GPS and regular WiFi, which is mainly used to connect with mobile devices installed on the two-wheeler or carried by the user(s). These modules are connected with an appropriate antenna. The main board is powered by a Lithium battery, given that the two-wheel vehicle could not include a power supply (e.g., bikes). A basic human-machine interface (HMI) is given by an extra board including LEDs and a Buzzer, which can be used to execute a basic services based on warnings or for testing purposes.

A reference prototype of the unit design has been implemented and installed in both a bike and a moped. Figure 2 shows the main parts of the hardware. An overall view of the platform is included in Figure 2a, where both the main unit (bottom in black) and the HMI board (upper in blue) are mounted on an electric moped. Figure 2b shows the hardware included in the main unit, which includes the functional modules of the MPU board described above. The unit is based on the Laguna LGN-20 platform from Commsignia. The communication board is visible on the top, from where different cables are connected to the 802.11p, regular WiFi (802.11n), GPS and 3G antennas. This system mounts an ARM11 300MHz SoC processor, 16 MB of Flash and 8 GB of internal storage, and 256 MB of RAM memory. The USB interface is used to connect with the HMI board. The power supply used is a 15 volts and 3500 mA lithium battery, which is able to maintain the unit up more than six hours in the operation modes used in the tests presented in Section V. The 802.11p antenna used is a 5.9 Ghz Taoglas Limited DCP.5900.12.4.A.02 (6 dBi), which has been affixed to the inner part of the unit enclosure, and it is visible on the upper right corner in Figure 2b. No 802.11n antennas have been used for the moment, and the cables are maintained in a foam piece, but communication with near devices has been possible without them. The 3G antenna is a common stick used for WiFi, and it is connected on the lower left corner of the unit.

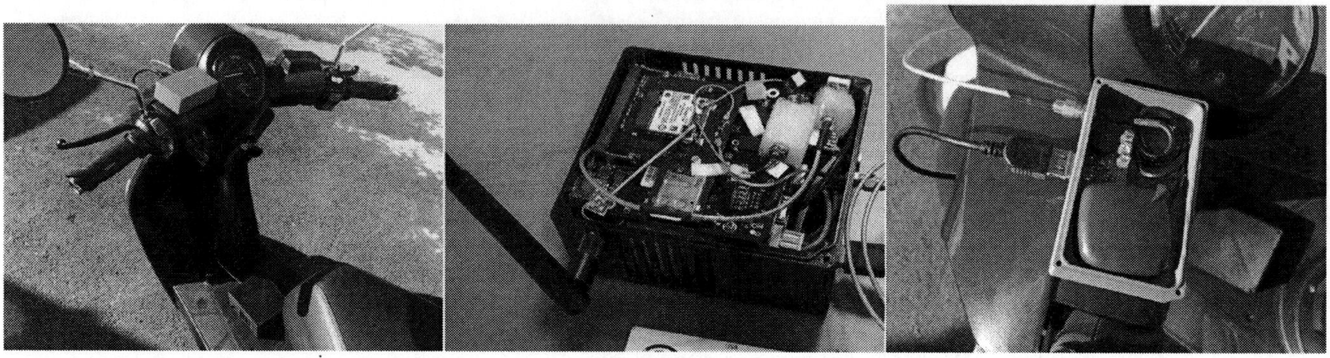

(a) Unit mounted on a moped (b) MPU board (c) HMI board

Figure 2. Prototype of the communication unit for two-wheel vehicles

The HMI board includes a set of LEDs and a buzzer. The prototype is showed in Figure 2c. One of the LEDs is used to inform about the whole unit status, while the others and the buzzer are left for application purposes, such as the safety service for two-wheelers described in [19]. The USB connector is used to connect the circuit to the main board, as can be seen in Figure 2a. The GPS antenna has been finally installed in the same enclosure used for the HMI in order to avoid interferences with the main unit electronics and improve the signal reception. This antenna is a PCTEL WS3917, which has worked correctly in the tests.

IV. COMMUNICATION STACK

The communication stack of the new embedded unit for two-wheelers has been ported from the one used in the car mobile router presented in [20]. This stack follows the ISO/ETSI reference architecture specifications [21][22]. Its main design blocks are depicted in Figure 3. IPv6 connectivity is supported by the set of elements included within the networking and transport layer of the unit. Network Mobility Basic Support (NEMO) [23] is in charge of maintaining IPv6 reachability. Regarding security, the mobile router is equipped with the needed elements to secure mobility-related traffic by means of Internet Protocol Security (IPsec) [24]. Communication flows can be secured with IPsec once security associations are stablished with Internet Key Exchange Protocol Version 2 (IKEv2) [25].

The functionality provided by NEMO is useful to maintain Internet connectivity in C-ITS between all the nodes mounted on the vehicle and the infrastructure. Thanks to NEMO, mobile devices connected to the mobile router of the two-wheeler (e.g., a mobile phone), are reachable through the infrastructure at the same IPv6 address during the itinerancy of the bike or moped. Additionally, with the aim of supporting multihoming, Multiple Care-of Addresses Registration (MCoA) [26] has been included in our unit. This technology allows us to perform faster handovers, maintaining initial and target communication flows up during the transition. However, in order to better decide the most suitable network at every single location, we have also included modules from the IEEE 802.21 standard. As discussed in [20], with this technology it is possible to obtain seamless handovers by minimizing packet losses during the process.

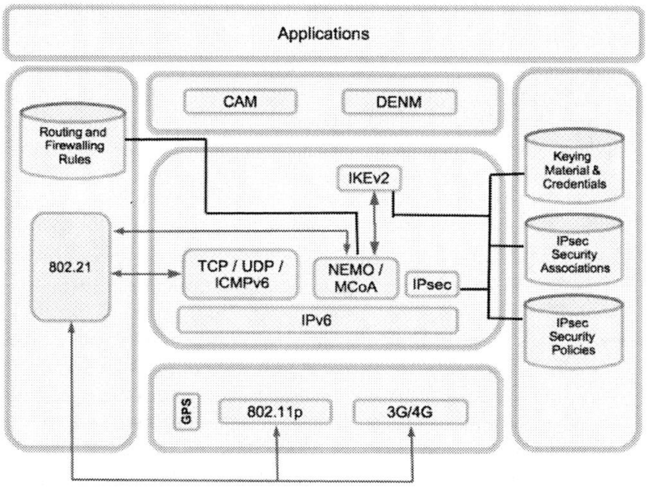

Figure 3. Design of the mobile router with extended IPv6 mobility support

CAM [27] and DENM [27] messages are supported by the communication stack. They can be used to develop ITS applications, such as the safety application for two-wheelers presented in [19]. In this work, CAM messages are encapsulated in UDP datagrams over IPv6, which are sent to the all-nodes multicast IPv6 address in a direct V2V fashion.

V. EXPERIMENTAL EVALUATION

The communication unit for two-wheelers has been tested to assess its performance in real driving settings. Given its relevance for vehicular applications, the evaluation has been focused on the 802.11p channel, through a set of one-hop tests.

A. Testbed

The testing scenario was set in the surroundings of the University Centre of Defence at the Spanish Air Force Academy. For the sake of simplicity, we preferred to move a mobile router in a car, and maintain static the two-wheeler node, since the last one has a battery and it is not necessary an external power supply. Figure 4a shows the open road where the tests were performed, in which there is direct line of sight between

(a) Testing location (b) Two-wheeler communication node (c) Mobile router used in the car

Figure 4. Scenario for testing one-hop communications using 802.11p

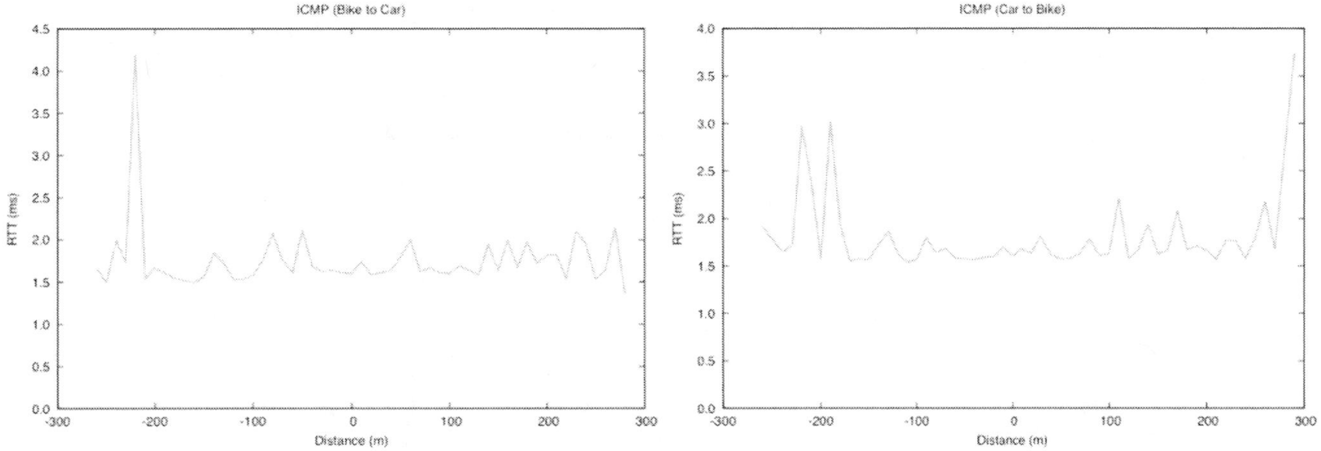

(a) Two-wheeler to car (b) Car to two-wheeler

Figure 5. Delay tests using one-hop communication with 802.11p

the two-wheeler unit and the car mobile router. The new embedded node was detached from the moped and placed in a box on the sidewalk, next to the road and elevated 10 cm, as can be seen in Figure 4b. The HMI board (blue box) was connected just to check the correct operation of the unit through one of the LEDs, while the GPS antenna was necessary to geo-localise the unit. The box was placed at the middle of the road stretch showed in Figure 4a. The mobile router mounted on a common car is showed in Figure 4c. This is a Laguna LGN-20 from Commsignia, running a communication stack equivalent to the one included in the two-wheeler node. The antenna used is a combined omnidirectional 3G/11p/GPS 7dBi, which was attached on the vehicle roof with a magnetic base.

The tests have been carried out with three different protocols:

- ICMPv6, to check the delay of the communication link. A continuous check of the link using the *ping6* tool at a rate of 1 Hz and with a payload of 56 bytes has been performed.
- UDP, to study packet losses. It has been chosen a transmission rate of 1 Mbps, sending 1230 bytes of data in each datagram. The *iperf* tool has been used to generate this UDP traffic.
- TCP, to study the maximum achievable throughput. The *iperf* tool has been used for this again.

Each transmission type was used to pass six times with the car near the two-wheeler node, and different tests were carried out to check the car to two-wheeler transmission direction, and the two-wheeler to car one. Each record of the test was geo-located by marking it with the GPS position. This way the results obtained have been averaged in a 10 meter basis, computing the distance from the car to the two-wheeler node.

B. Results

The delay results obtained in the tests are showed in Figure 5. As can be seen, a good performance is obtained within a communication range of near 600 meters. The round-trip time value obtained in most of the stretch is between 1.5 and 2 ms. Delay peaks are obtained at the edges of the road stretch until the communication is broken. As can be seen in the plots, the results gathered in both communication directions are equivalent, given that the *ping6* generates packets that reach the destination and then come back.

The study of the packet delivery ratio (PDR), which is measured in percentage of packets that reach the destination successfully, is showed in Figure 6. As expected, a similar communication range is obtained here and, although the distance with the two-wheeler node affect the performance at

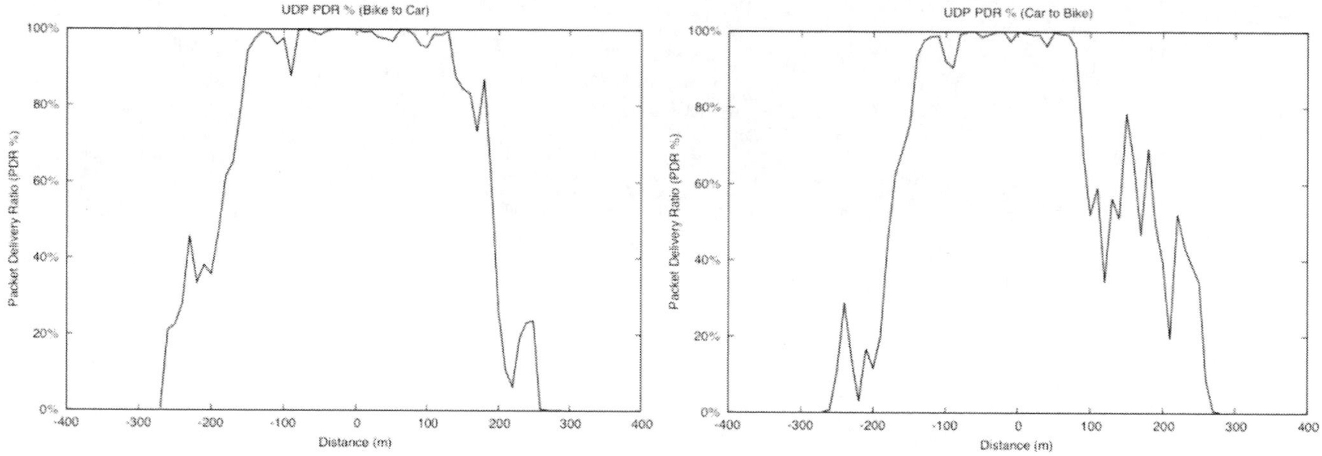

(a) Two-wheeler to car

(b) Car to two-wheeler

Figure 6. Packet delivery tests using one-hop communication with 802.11p

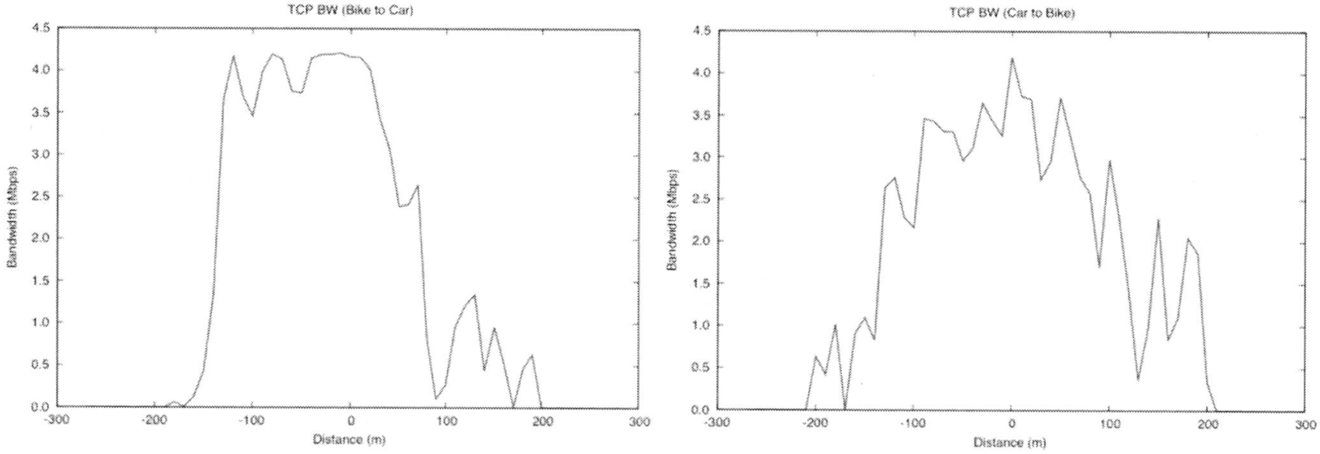

(a) Two-wheeler to car

(b) Car to two-wheeler

Figure 7. Throughput tests using one-hop communication with 802.11p

distant points, good PDR results are obtained in general within the road stretch. It can be observed that the two-wheeler to car case performs better. This is due to the better reception sensibility of the antenna used for the case of the car.

TCP results are plotted in the two graphs included in Figure 7. The first noticeable difference with the previous tests is the communication range decrease. This is attributed to the features of the communication protocol used, since TCP needs a successful connection establishment stage prior to start the transmission. After that, the transmission rate is adapted according to the detected performance of the link. Again, it is observed that in the two-wheeler to car case, the link performs in a steadier way. In any case, the maximum throughput obtained reaches 4 Mbps in both transmission directions, which is a good value.

The results obtained indicate a good performance of the 802.11p communication link. The delay and PDR results assure a good operation of the network for safety services requiring direct V2V communications. Moreover, the good throughput of the link enables the cellular network offload for services, such as video transmission or file downloading, always when a near roadside unit is available. Nevertheless, it must be considered that, first, the expected performance when driving near multiple vehicles using 802.11p would be impacted by the congestion of the communication channel; and, second, urban scenarios would imply additional signal reception issues, due to the rest of vehicles and buildings.

VI. CONCLUSION

The paper describes the work carried out to develop a communication unit for two-wheel vehicles and its evaluation under real settings. The design of the unit has been adapted to the distinguishing features of two-wheelers, such as the need of a battery and the space/interface limitations. The hardware provides cellular and short-range communications, while a proper middleware has been added to support IPv6 networking. Both Internet-based and direct V2V communications are possible and, due to the inclusion of IPv6 protocols, the unit is ready for novel Future Internet environments, such as smart cities.

The performance tests carried out with short-range communications demonstrate the capabilities of the unit for connecting with road-side units or nearby vehicles. The results obtained indicate that the communication channel presents an RTT delay of 2 ms for delivery packets, a PDR above 50% within a road stretch of 400 meters, and a maximum throughput of 4 Mbps. Given the embedded design of the new unit, these are good results that assure the operation of the unit for a number of potential services.

At the moment, we are further evaluating the unit considering the network mobility capabilities, and our plans consider the integral adoption of all means of transport in smart environments powered by IoT technologies. For this to be done, especial IoT protocols will be adapted for the vehicle domain and research efforts are being identified for homogenizing data recovery and processing.

ACKNOWLEDGMENTS

This work has been sponsored by the Spanish Ministry of Economy and Competitiveness through the EDISON project (contract TIN2014-52099-R), and the Spanish Ministry of Home Affairs - Department of Transport (DGT) through the S-CICLO project (SPIP2015-01757).

REFERENCES

[1] "Connecting Vehicles for Safe, Confortable and Green Driving on European Roads (Drive C2X)," http://www.drive-c2x.eu, 2013, [retrieved: June, 2017].

[2] "European Field Operational Test on Safe, Intelligent and Sustainable Road Operation (FOTsis)," http://www.fotsis.com, 2015, [retrieved: June, 2017].

[3] USA Department of Transportation, "Federal Motor Vehicle Safety Standards: Vehicle-to-Vehicle (V2V) Communications," USA Department of Transportation, August 2014.

[4] European Commission, "Directive 2010/40EU of the European Parliament and of the Council of 7 July 2010 on the framework for the deployment of Intelligent Transport Systems in the field of road transport and for interfaces with other modes of transport," European Commission, August 2010.

[5] C. Englund, L. Chen, A. Vinel, and S. Yang, "Future applications of VANETs," in Vehicular ad hoc Networks: Standards, Solutions, and Research, C. Campolo, A. Molinaro, and R. Scopigno, Eds. Berlin: Springer, 2015, ch. 18, pp. 525–544.

[6] T. Gandhi and M. Trivedi, "Pedestrian protection systems: Issues, survey, and challenges," Intelligent Transportation Systems, IEEE Transactions on, vol. 8, no. 3, September 2007, pp. 413–430.

[7] C. Sugimoto, Y. Nakamura, and T. Hashimoto, "Prototype of pedestrian-to-vehicle communication system for the prevention of pedestrian accidents using both 3G wireless and WLAN communication," in Wireless Pervasive Computing, 2008. ISWPC 2008. 3rd International Symposium on, May 2008, pp. 764–767.

[8] K. David and A. Flach, "CAR-2-X and pedestrian safety," IEEE Vehicular Technology Magazine, vol. 5, no. 1, March 2010, pp. 70–76.

[9] W. Jin, C. Kwan, Z. Sun, H. Yang, and Q. Gan, "SPIVC: A smartphone-based inter-vehicle communication system," in Transportation Research Board 91st Annual Meeting 2012, January 2012, pp. 1–12.

[10] M. Nagai, K. Nakaoka, and Y. Doi, "Pedestrian-to-vehicle communication access method and field test results," in Antennas and Propagation (ISAP), 2012 International Symposium on, Oct 2012, pp. 712–715.

[11] X. Wu, S. Subramanian, R. Guha, R. G. White, J. Li, K. W. Lu, A. Bucceri, and T. Zhang, "Vehicular communications using DSRC: Challenges, enhancements, and evolution," IEEE Journal on Selected Areas in Communications, vol. 31, no. 9, September 2013, pp. 399–408.

[12] S. Cespedes, J. Salamanca, J. C. Sacanamboy, and C. Rivera, "Poster: Smart networked bicycles with platoon cooperation," in 2014 IEEE Vehicular Networking Conference (VNC), Dec 2014, pp. 199–200.

[13] R. Miucic, S. Rajab, S. Bai, J. Sayer, and D. Funkhouser, "Improving motorcycle safety through DSRC motorcycle-to-vehicle communication," in SAE 2015 World Congress and Exhibition, April 2015, pp. 1–8.

[14] U. Hernandez-Jayo, J. Perez, and I. de-la Iglesia, "Poster: Wearable warning system for improving cyclists safety in the scope of cooperative systems," in Vehicular Networking Conference (VNC), 2015 IEEE, Dec 2015, pp. 153–154.

[15] U. Hernandez-Jayo, I. De-la Iglesia, and J. Perez, "V-alert: Description and validation of a vulnerable road user alert system in the framework of a smart city," Sensors, vol. 15, no. 8, 2015, pp. 18 480–18 505.

[16] K. Dhondge, S. Song, B. Y. Choi, and H. Park, "Wifihonk: Smartphone-based beacon stuffed WiFi car2x-communication system for vulnerable road user safety," in 2014 IEEE 79th Vehicular Technology Conference (VTC Spring), May 2014, pp. 1–5.

[17] J. J. Anaya, E. Talavera, D. Gimnez, N. Gmez, J. Felipe, and J. E. Naranjo, "Vulnerable road users detection using v2x communications," in 2015 IEEE 18th International Conference on Intelligent Transportation Systems, Sept 2015, pp. 107–112.

[18] A. Dhraief, N. Montavont, R. Kuntz, and M. Tsukada, "E-bicycle demonstration on the tour de france," in Computing in the Global Information Technology, 2007. ICCGI 2007. International Multi-Conference on, March 2007, pp. 56–56.

[19] J. Santa, P. J. Fernandez, and M. A. Zamora, "Cooperative its for two-wheel vehicles to improve safety on roads," in 2016 IEEE Vehicular Networking Conference (VNC), Dec 2016, pp. 254–257.

[20] P. J. Fernandez, J. Santa, F. Pereniguez, and A. F. Skarmeta, "Towards seamless inter-technology handovers in vehicular {IPv6} communications," Computer Standards & Interfaces, vol. 52, 2017, pp. 85 – 96.

[21] ETSI TC ITS, "Intelligent Transport Systems (ITS); Communications Architecture," ETSI EN 302 665, European Telecommunications Standards Institute, September 2010.

[22] ISO TC 204, "Intelligent transport systems - Communications Access for Land Mobiles (CALM) - Architecture," ISO 21217, International Organization for Standardization, april 2013.

[23] V. Devarapalli, R. Wakikawa, A. Petrescu, and P. Thubert, "Network Mobility (NEMO) Basic Support Protocol," RFC 3963 (Proposed Standard), Jan. 2005.

[24] S. Kent and K. Seo, "Security Architecture for the Internet Protocol," RFC 4301 (Draft Standard), Internet Engineering Task Force, Dec. 2005.

[25] C. Kaufman, P. Hoffman, Y. Nir, P. Eronen, and T. Kivinen, "Internet Key Exchange Protocol Version 2 (IKEv2)," RFC 7296 (Proposed Standard), Internet Engineering Task Force, Oct. 2014.

[26] R. Wakikawa, V. Devarapalli, G. Tsirtsis, T. Ernst, and K. Nagami, "Multiple Care-of Addresses Registration," RFC 5648 (Proposed Standard), Internet Engineering Task Force, Oct. 2009.

[27] ETSI TC ITS, "Intelligent Transport Systems (ITS); Vehicular Communications; Basic Set of Applications; Part 2: Specification of Cooperative Awareness Basic Service," ETSI EN 302 637 2, European Telecommunications Standards Institute, November 2014.

A Method of Detecting Camouflage Data with Mutual Vehicle Position Monitoring

Shuntaro Azuma
Graduate School of Science
and Engineering Information Engineering
Doshisha University
Kyoto, Japan
email:syuntaro.azuma@nislab.doshisha.ac.jp

Manabu Tsukada
Graduate School of Information
Science and Technology
Tokyo University
Tokyo, Japan

Teruaki Nomura and Kenya Sato
Graduate School of Science
and Engineering Information Engineering
Doshisha University
Kyoto, Japan

***Abstract*—Due to the development of V2V communication, such safe driving support as collision prevention and adaptive cruise control has been achieved. Furthermore, in recent years, in addition to infrastructure-to-vehicle communication (V2I communication) and communication with a cloud server using mobile lines is also possible (V2C communication), and such communication is generally called V2X communication. Through V2X communication, vehicle's peripheral information can be shared with other vehicles on a cloud server. However, the influence of inappropriate information on the cloud must be addressed. By faking vehicle information, a system using a cloud server, perhaps deliberately causing congestion and/or accidents. In this research, we propose a method that detects camouflage data from all of that aggregated data on a cloud server using V2X communication and utilizing the surrounding vehicle information. We also analyze possible threats and the requirements for the data that are sent to a cloud, clarify security, and evaluate the proposed method's implementation. We detected 93% of the camouflage data, and improved the detection rate 100% by increasing the threshold value of the proposed method and, enhancing the effect of guaranteeing the data's reliability. Furthermore, we showed the false positives of the proposed method and its execution processing time and examined feasibility.**

Keywords–vehicle security; V2X communication; detecting camouflage data.

I. INTRODUCTION

In recent years, research on automatic driving and V2V communication is being conducted in the Intelligent Transport Systems (ITS) field. In addition to providing V2V communication using the Vehicular Ad hoc Network (VANET), vehicles can engage in V2I communication with roadside aircraft and V2P communication with tablets owned by pedestrians. Vehicles can also do V2C communication with a cloud server using mobile lines, and the kinds of communication we mention here are generally referred to as V2X communication. While vehicles perform V2X communication, a cloud server can collect various kinds of information, and we can create a Local Dynamic Map (LDM) [1] for cooperative driving from the collective management of road and vehicle information. This type of communication sometimes is referred as probe information systems [2] or floating car data (FCD) [3]. In addition, various systems and services can be provided, such as the simplification of such management tasks as summarizing operation results, analyzing operation trends, summing up tasks, and simplifying the input of daily reports.

On the other hand, in a system using a cloud server, camouflage data transfer to a cloud influences a system. Attacks against safe driving support services using a cloud also pose a threat because the intentional transfer of camouflage data to a cloud camouflage acts are on the rise. Attackers can block roads or cause traffic congestion by sending to a cloud a camouflage information that pretends to be involved in an accident. As a type of vehicle disguising acts, various camouflage acts have been identified, such as camouflaging both driving and position information as well as the vehicle's condition. In this research, we focus on camouflage position information among all of the data received by a cloud from vehicles and detect them by mutually monitoring the position information of vehicles using V2X communication.

II. ANALYSIS THREAT OF TRANSMISSION DATA

There are things researching the detection of malicious vehicles in V2X communication [4] [5], but in reality malicious vehicles are ambiguous. In this section, we analyze attacks on vehicle communication and show what kind of malicious vehicles to be solved in this research.

A. *Threats, Requirements, and Resolution example*

Table I shows analysis of the transmission data to a cloud server. The threats include eavesdropping attacks, falsifications, and spoofing. Spoofing attacks are divided into vehicle pretense and data camouflage. Vehicle pretense means that attackers pretend to be other vehicles. For example, even though one vehicle doesn't have any trouble, an attacker pretends to be the vehicle and then calls the police with a lie that it had an accident. An example of data camouflage is when a vehicle's own position information or status is masked.

Security requirements about threats includes confidentiality, completeness, node reliability, and data reliability. To supply confidentiality and completeness, data encryption is proposed and can be done by a secret key or an ID base cipher. Node reliability identifies vehicles that are pretending to be other vehicles. The Public Key Infrastructure (PKI) method, which is adapted by the vehicles, is one good resolution

TABLE I. ANALYSIS THREATS ABOUT TRANSMISSION DATA

THREAT		REQUIREMENT	COUNTERMEASURE
Eavesdropping		Confidentiality	Encryption
Falsification		Completeness	Encryption
Spoofing	Vehicle pretense	Node reliability	PKI
	Data camouflaging	Date reliability	**Target of this research**

Figure 1. PKI to adapt to vehicles

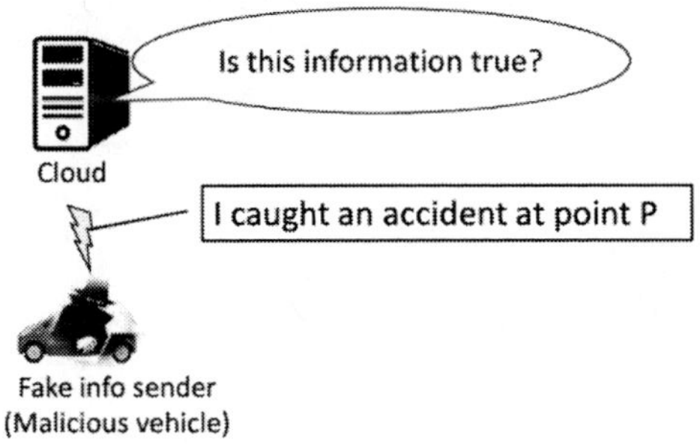

Figure 2. Problem of settling by this research

because the certificate guarantees vehicles. Data reliability prevents attackers from camouflaging data. It doesn't involve resolution at all.

B. Difference Between Node and Data Reliability

Node reliability means that a cloud server trusts a particular vehicle and believes that it isn't pretending to be a different vehicle. The previous section showed that, PKI measures adapts to the vehicles to resolve the problem. A cloud might be able to verify the electronic certification and confirm the transmitter information by the mechanism shown in Figure 1.

This research focuses on such data camouflages, as described in Figure 2 in spoofing acts. Since the data encryption and PKI don't confirm whether the received data are camouflage, the problem, which is resolved by these methods, is different from the one we focus on in this research. We propose a method that can handle such examples as when the given matter, which guarantees the reliability of the data in an act, camouflages vehicle data.

III. PROPOSAL

In this section, we propose a method to detect camouflage data from among data sent to the cloud.

A. Outline

Vehicles use V2X communication. When they send their position information to a cloud server, they also send other information than just their position. In this research, the cloud detects camouflage data from the received data using the relay base station information in V2C communication and peripheral vehicles in V2V communication. We explain them separately to simplify the movement outline of proposed technique.

B. Presuppositions

1) A safe channel has been secured by a preliminary relationship of mutual trust among all vehicles and the cloud server
2) Vehicles and the cloud have been mutually certified beforehand.
3) Relationships of mutual trust have been built by a base station with a cloud server.

C. Definition of Terminology in Proposed Method

- Vehicle ID

The ID used by a vehicle during V2V communication; a public ID that is different for every vehicle.

- V2C Vehicle ID

The ID used for peculiar questions during V2C communication. This secret ID is not available to others. V2C Vehicle and Vehicle IDs are uniquely related.

- Via Base Station (BS) ID

This unique ID in a relay base station for V2C communication. It adopts bidirectional one time ID signaling, and establishes its life-time.

- Peripheral Vehicle (PV) ID

The ID received by a vehicle from other vehicles by V2V communication.

D. Use of Base Station Information in V2C Communication

When sending position information using V2C communication, a vehicle attaches the V2CVehicleID to its position information and sends it to the cloud. The relay base station on the V2C communication, is encapsulated and make a header from the ViaBSID in the position that was sent from a vehicle. The V2CVehicleID for all of the vehicles preserve the registration beforehand in the cloud, which knows a request that can confirm vehicle's identify by checking the V2CVehicleID. The possible communication range covered by a base station's area and the ViaBSIDs are also registered with cloud.

Figure 3 indicates an example of base station information in V2C communication. The vehicles possess V2CVehicleID; the base stations possess ViaBSIDs. The V2CVehicleIDs are regarded as either V2C_A or V2C_B, and ViaBSID, where the base stations are unique, is made by BS1 or BS2 as simple explanations. When a vehicle performs V2C communication, it obtains the information that the cloud will a relay base station as well as its position information: ViaBSID and V2CVehicleID.

Figure 4 shows a countermeasure example of a position data camouflage act. We can detect the position information that is being camouflaged at another base department using the relay base station information in V2C communication.

E. Using Peripheral Vehicle Information in V2V Communication

Vehicles exchange VehicleIDs with nearby vehicles using V2V communication. A vehicle views the other vehicles other vehicles in the potential V2V communication area as peripheral vehicles. A VehicleID that is received from a peripheral

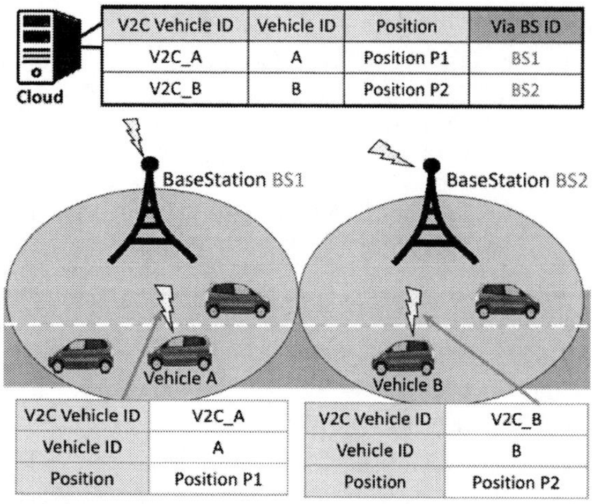

Figure 3. Use example of base station information in V2C communication

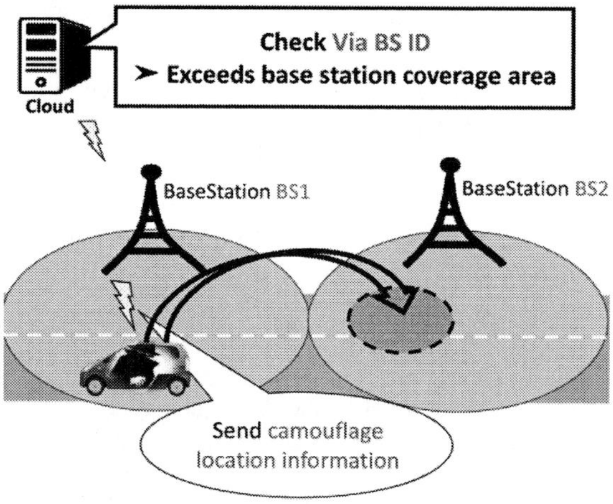

Figure 4. Advantage of using base station information

vehicle is regarded as peripheral vehicle information (PVID). In our proposed technique, only VehicleID information is exchanged by V2V communication.

When a vehicle sends its position information to a cloud, the V2CVehicleID, its VehicleID, and the PVID are attached by V2V communication. The PVID shows a guarantee that nearby vehicles are in the V2V communication possible area. Figure 5 shows an example of peripheral vehicle information by V2V communication. Vehicle A communicates with the others vehicles in the area where V2V communication is possible and acquires VehicleIDs from Vehicles C and D. Vehicle A handles the acquired VehicleIDs as PVIDs and verifies their nearby position with peripheral Vehicles C and D.

Figure 6 shows a countermeasure example of a position data camouflage act. We assume that a malicious vehicle camouflaged its position information. The cloud confirms the PVID sent with the position information from a vehicle, and compares a vehicle's position information that is relevant to the PVID with the information of the vehicle that transmitted. When comparing the position information that is outside the possible V2V communication area, the cloud determined that the received position information has been camouflaged. However when the position information does

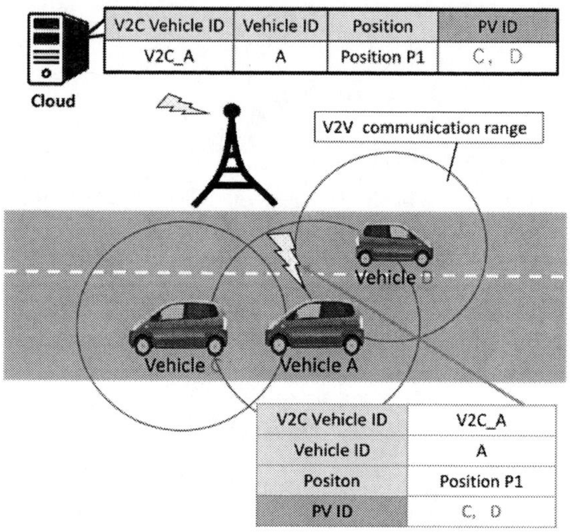

Figure 5. Use example of peripheral vehicle information in V2V communication

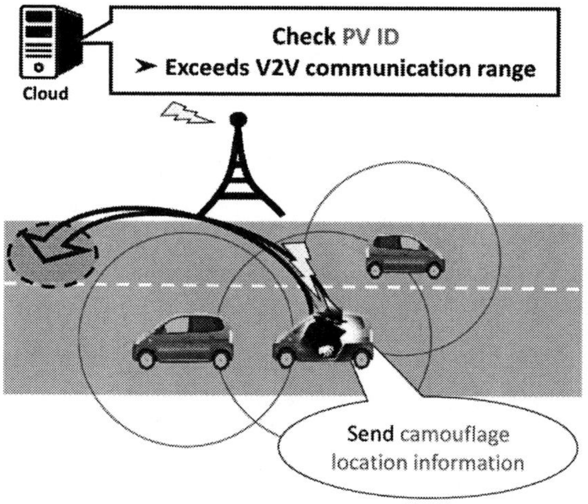

Figure 6. Advantage of using peripheral vehicle information

not exceed the area, the cloud trusts the received position information. Vehicles acquire peripheral vehicle information using V2V communication and mutually monitor it. This helps the cloud detect camouflage data.

F. Detection Method of Camouflage Data With V2X Communication

Our proposed technique is a combination of the two described above by using V2X communication (Figure 7). The cloud receives not only the position and the VehicleID information data but also the data attached to the peripheral vehicles and the relay base station information monitored mutually by V2X communication. Camouflage data can be detected through these data, as described in Figure 8.

The V2CVehicleID is used in the first step on Figure 8. Next, we verified whether the data received by the cloud were sent from a vehicle. Second, we compared the ViaBSID with the position information of a transmission vehicle to confirm whether it exists in the area covered by the relay base station. When the position information of the transmission vehicle

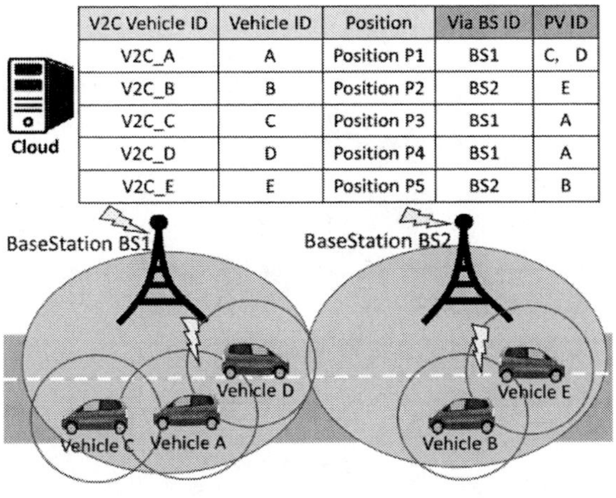

Figure 7. Use example of peripheral vehicle information in V2X communication

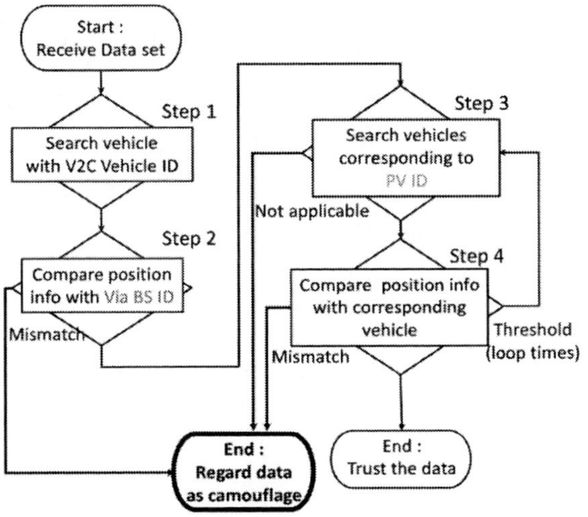

Figure 8. Camouflage data detection procedure to a vehicle send data

exceeds the area covered by the relay base station, we assume that consistency can't be maintained, and that the position information was authorized by camouflage data. A camouflage act in another base department is stopped by this step. In the third or fourth stage, camouflage data are detected based on vehicle information near the transmission vehicle using a PVID. The third step's adjustment-lessness occurs when a vehicle, which doesn't exist around the transmitter, is regarded as a peripheral vehicle. It isn't possible to camouflage another nearby vehicle because a mutual VehicleID exchanged for peripheral vehicles by the V2V communication. The position information of the data, which didn't reach the prescriptive number of times, or peripheral vehicles and the data which couldn't maintain consistency, are identified as camouflage at the fourth stage. We can treat a location's camouflage beyond the possible V2V communication area and make the data more credible by setting up specified execution count. The data, which didn't meet the condition, violates authorization and can't be trusted in the second, third, or fourth stages.

IV. EVALUATION AND CONSIDERATION

To evaluate the usefulness of our proposed camouflage data detection method, we calculate the evaluation. And then we consider the practicality of our proposal from the evaluation obtained.

TABLE II. SIMULATION PARAMETER

Simulator	Scenargie2.0	
Vehicle number	158 [cars] (five of the send camouflage positions.)	
Area	1000 [m] × 1000 [m]	
Communication mode	ARIB STD T109	LTE
Use frequency band	700 [Mhz]	2.5 [GHz]
Communication interval	100 [ms]	1.0 [s]
Radio spread model	ITU-R P.1411	LTE-Macro
Base station ground clearance	1.5 [m]	

A. *Simulator*

In this paper, we used Scenargie [6] as a simulator to evaluate the performance of our evaluation of the proposed method. Scenargie is a network simulator developed by Space-Time Engineering (STE). By combining expansion modules, like LTE, V2V communication and, multi-agent simulation can be constructed. In addition, since communication systems and evaluation scenarios are becoming more complicated, this ingenious simulation has greatly reduced the effort required to create scenarios. Examples include GUI scenario creation, map data, the graphical information display of a communication system, and a radio wave propagation analysis function.

B. *Evaluation Model*

For the evaluation environment, we used a 1 square kilometer square Manhattan model and the simulation parameters shown in Table II. We set the number of vehicles to 158 [cars] and the range to 1 [km^2] because the average car density across Japan is 158 [cars/km^2]. The ITU-R P.1411 model is a radio wave propagation scheme that considers road map information, and radio waves are attenuated based on the shape of the road, so we compared our model with a two-ray model using direct waves and reflected waves from the ground. This model closely resembles reality.

C. *Evaluation of Camouflage Data Detection*

Figure 9 shows the per-threshold detection rate of the camouflaged data from data aggregated in a cloud server. A camouflage data transmitted to a cloud could be detected at 100% by increasing the proposed method's threshold. However, when the threshold was low, completely detecting all of the incorrect data was impossible. The reason is shown in Figs. 10 and 11. The former shows an example where location information can be disguised in a possible range within a range in which peripheral vehicles and the V2V communication range are possible. In this case, since peripheral vehicles guarantee camouflage information from a malicious vehicle, camouflaging the position information becomes possible. Figure 11 show a collusion between malicious vehicles. Since they guarantee mutual position camouflage information, there are trying to fool the cloud into trusting the camouflage data. In other words, this is an inadequate PVID to help with data spoofing. These problems can be addressed by increasing the prescribed number of times (threshold values) shown in Figure 8. By increasing the threshold values, we can create the situation shown in Figure 12 and it is possible to limit deception by malicious vehicles.

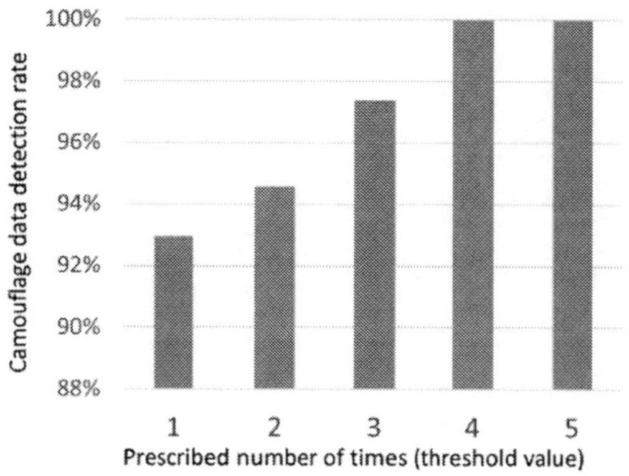

Figure 9. Detect rate of the camouflage data in cloud concentration data

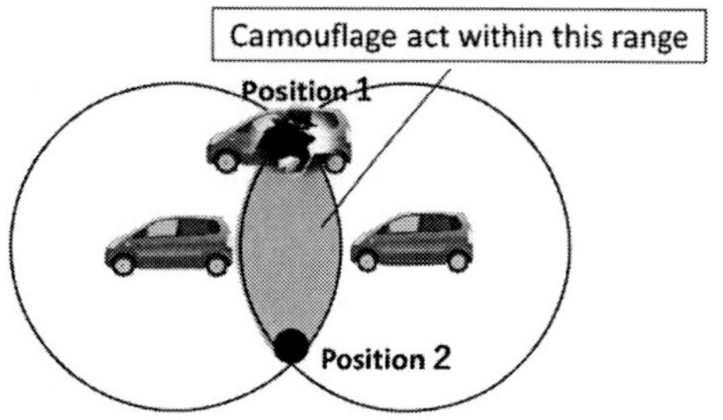

Figure 10. Camouflage acts in V2V communication coverage with peripheral vehicles

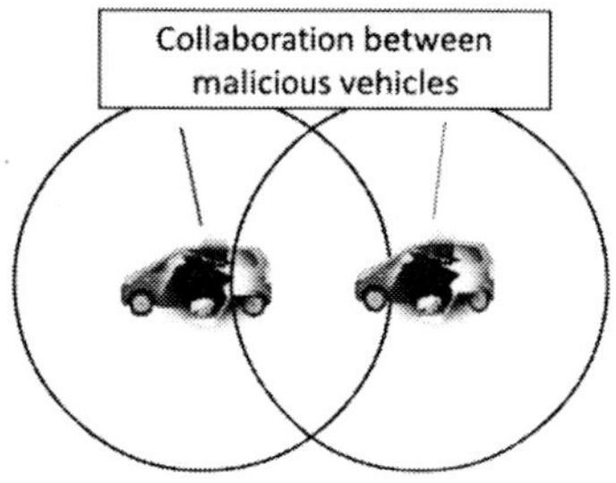

Figure 11. Collusion between malicious vehicles

D. *Evaluation of Misdetection Rate*

Figure 13 shows the false detection rate (false positives) of the proposed method based on the average car density in Japan. The method's threshold is the amount of information data of the peripheral vehicles that is necessary for a cloud to trust the information. In the previous section, we found that an increase in the threshold improves the detection rate of the camouflage data. Here, we consider the false positive detection rate, (false positive), regarding whether a cloud trusts information on vehicles that are not conducting camouflage activities. In the simulation environment shown in Table II, Figure 13 shows that not all 158 cars are doing camouflage acts and the false positives were measured. By increasing the threshold value, the false detection rate improved. Increasing the threshold value in the average Japanese vehicle density erroneously detects normal communication as abnormal.

Therefore, the false positives under the average vehicle density environment in Osaka, which has the highest average car density in Japan, are indicated by Figure 14. In a high vehicle density zone, since much peripheral vehicle information can be acquired by V2V communication, even if the threshold is increased, an increase in the false detection rate can be suppressed. Therefore, we found that the proposed method is more effective in areas with high vehicle density. Actually, the influence of camouflage acts of vehicle information is great in areas with high vehicle density. The proposed method, can guarantee that the information transmitted by vehicle to a cloud is better in areas where more peripheral vehicles exist than in areas with fewer peripheral vehicles. Our proposed method is

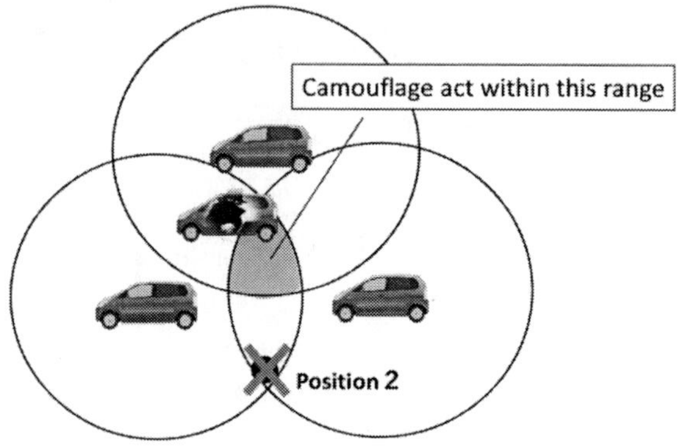

Figure 12. Restriction on camouflage acts accompanying increase in information on peripheral vehicles

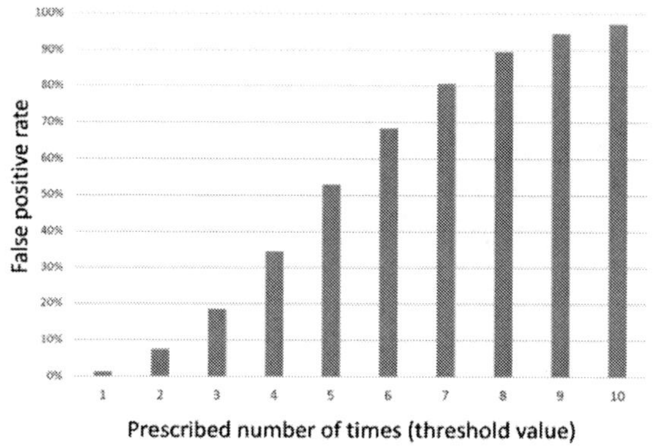

Figure 13. False positives by threshold value under Japanese average vehicle density (158[cars/km^2]) environment

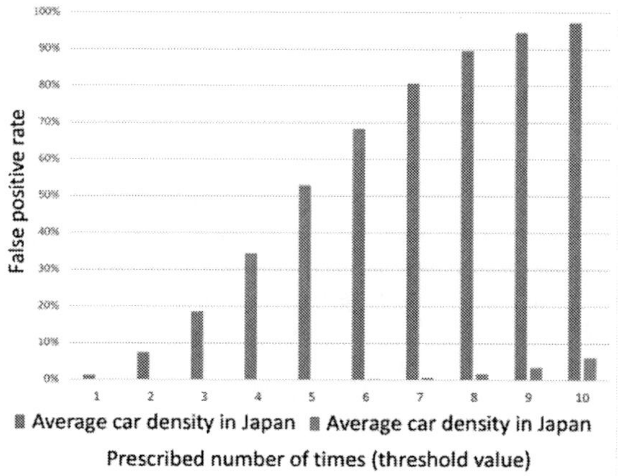

Figure 14. False positive comparison with Osaka average vehicle density (1128 [cars/km^2]) environment

TABLE III. ENVIRONMENT IN THE PROCESSING TIME MEASUREMENT

OS	macOS Sierra
Processor	1.6GHz Intel corei5
Memory	8GB 1600MHzDDR3
Script	Python
Data base	MySQL

TABLE IV. PROCESSING TIME OF UNJUST MEASURE TO A VEHICLE OF SEND DATA

Threshold	Detected in step2	Detected in Figure 8's step4	Usual end
1	0.1[ms]	0.31[ms]	0.31[ms]
2	0.1	[0.31,0.53]	0.53
3	0.1	[0.31,0.76]	0.76
4	0.1	[0.31,0.96]	0.96
5	0.1	[0.31,1.2]	1.2
6	0.1	[0.31,1.4]	1.4
7	0.1	[0.31,1.6]	1.6
8	0.1	[0.31,1.8]	1.8
9	0.1	[0.31,2.0]	2.0
10	0.1	[0.31,2.2]	2.2

useful in traffic congestion zones where self-vehicle spoofing acts have a huge impact.

E. Measurement of Processing Time

In the evaluation environment shown in Table III, the processing time necessary for the detection of camouflage data is evaluated by Table IV, based on the detection method of camouflage data in Figure 8. The processing time of one vehicle is shown. By using BSIDs, camouflage data can be detected at the beginning of the processing by the proposed method, and the processing time becomes relatively fast. In the detection method using PVIDs, the processing time is different for each threshold. By increasing the threshold value, the detection procedure of camouflage data by PVID is repeated. Even during the repetition, since the processing time changes depending on whether the comparative data can be found relatively early or in the final stage, a range was set for the processing time up to Step 4. A case where no camouflage data is not detected is defined as normal termination and the upper limit of the processing time at that threshold is indicated. As the threshold of the proposed method increases, the processing time required for normal termination increases. We must determine the threshold values based on the V2C communication delay and the allowable range of the delay times of safe driving support systems.

V. CONCLUSION

In the Intelligent Transport Systems (ITS), using a cloud server is inevitable. For providing a safe driving support service using a cloud, camouflaging vehicle information and spoofing a vehicle are threatening. In this research, we used V2X communication, obtained information from various objects, and described measures against vehicle spoofing. We proposed a method that detects camouflage data from the information transmitted by vehicles to a cloud. By using the information of a relay base station in V2C communication and the peripheral vehicle information using V2V communication, measures are taken against camouflaging vehicle information. By increasing the proposed method's threshold, the detection rate of camouflage data was improved and vehicle information was made more reliable. Our proposed method can be adapted to depopulated regions by changing the amount of the data of peripheral vehicle information required as the detection rate improves based on car density. In overcrowded vehicle ares, we confirmed that our proposed method is the most effective.

ACKNOWLEDGMENT

A part of this work was supported by KAKENHI (JP 16H02814) and MEXT program for the strategic research foundation at private universities.

REFERENCES

[1] Intelligent Transport Systems (ITS); Vehicular Communications; Basic Set of Applications; Local Dynamic Map (LDM), September 2014. ETSI EN 302 895 V1.1.1 (2014-09).

[2] International Standardization Organization. (2009). Vehicle probe data for wide area communications (ISO 22837:2009).

[3] Schafer, R. P., Thiessenhusen, K. U., Brockfeld, E., and Wagner, P. (2002). A traffic information system by means of real-time floating-car data.

[4] Yang, Yuchen Ou, Dongxiu Xue, Lixia Dong, and Decun, Infrastructure-based Detection Scheme of Malicious Vehicles for Urban Vehicular Network, Transportation Research Board 96th Annual Meeting, (2017).

[5] Gongjun Yan, Stephan Olariu, and Michele C. Weigle, Providing VANET security through active position detection, Computer Communications, (2008).

[6] " SPACE-TIME Engineering" , URL : https://www.spacetime-eng.com/en/ [accessed : 2017-07-07].

Positioning and Perception in Cooperative ITS Application Simulator

Ye Tao, Manabu Tsukada, Hiroshi Esaki
Graduate School of Information Science and Technology, The University of Tokyo
Email: {tydus, tsukada}@hongo.wide.ad.jp, hiroshi@wide.ad.jp

Abstract—**Intelligent Transportation System (ITS) is an intelligent system which can make the road traffic safer, more efficient, and more comfortable. The prerequisites of ITS are positioning, perception and networking. Different ITS infrastructures and applications are built by a large society from academia and industry. Nevertheless, the state-of-the-art network application simulators lack the capabilities to access the position and perception related information. In this paper, we proposed one extension to NS-3 network simulator and two to DCE application simulator, to bring the positioning and perception sensors to the simulated application. Both positioning and perception were designed fully experimental reproducible, and the simulated perception sensor can be adjusted to accommodate different types of devices. With the proposed work, much more scenarios of ITS applications can be tested, including Position Based Routing, Cooperative Awareness, etc.**

Keywords–Intelligent Transportation System; Vehicular Ad-hoc Network; Network Simulator; Network Application Simulator; Perception Aware Simulator

I. INTRODUCTION

Intelligent Transportation Systems (ITS) aim at optimization of the road traffic by realizing safe, efficient and comfortable transportation. Cooperative ITS is a branch of ITS features sharing the information between nodes to realize their common objects. Application of Cooperative ITS includes driver assistance in the near future, however the vehicular communication also remains essential in autonomous driving in order to support wider perception of the other vehicles around a vehicle that cannot be detected by the sensors equipped in the vehicle.

In order to connect among vehicles and roadside units, GeoNetworking [1] is employed as one of the network protocols in the ITS station architecture [2], because the geolocation based routing features the strength in the network with dynamic topology compared with topology based routing. GeoNetworking employs position-based routing VANET (PBR-VANET) to adapt to high-speed movement requirements. PBR-VANET is a type of VANET routing protocol that uses the position information of nodes to direct routing. It does not maintain routing tables or exchange link state information. Such a routing protocol shows better performance in a highly dynamic topology, where link states change frequently. However, the vehicle movement scenarios in the simulations are "hand-crafted" and not realistic enough nor scalable enough. We need a network simulator which can provide GPS function, to make the experiments scalable and realistic.

Cooperative Awareness (CA) Service is a service defined in the ITS station architecture [3]. It enables ITS vehicles to report their position through the GeoNetworking, making the automated driving software on the other ITS station be aware of the vehicle. Nevertheless, in the deployment period of ITS, we could not assume every vehicle to be ready for the CA service. Thus, some transition techniques will be required.

Contributions to CA service can be evaluated in the field operational tests (FOT). However, the FOTs can take very expensive devices and human power, which could render the experiments unscalable. If we come to a realistic perception-aware simulator, experiments in large fields and large scale network will become possible.

From the examples above, we know that conducting real field experiments are costly and not scalable. This is why a position and perception enabled simulator is essential for evaluating the ITS researches.

In the different fields, the researchers have invented different sorts of simulators, including:

Network Simulators models and simulates different kinds of network with differnt layers, from wireless radio to application protocols.

Program / Application Simulators provides reproducible and scalable simulations of real programs for programmer to debug.

Transportation Simulators understands real maps. It could simulate vehicle behaviors on the roads simultaneously and output their positions for the other simulators.

Different kinds of the simulators are not aware of the other parts. Network simulators do not know the road and vehicle information which came from the transportation simulators; vehicles in the transportation simulators cannot reflect the feedback from ITS application; application simulators are not capable of realizing the feedback from the vehicle simulator. This issue may be caused by the separation of different research fields: perception simulators are developed by the robotic researchers; transportation simulators are developed by vehicle researchers, etc. On the other hand, ITS applications may have all the portions above involved. Thus, it requires connections among the different simulators, to make an integrated simulated environment of the ITS application.

For our small step towards the integrated simulated environment, we propose a framework of realizing positioning and perception functionalities with the network and application simulators in this paper. It will enable the simulated applications to get the position of the vehicles they are deployed into, which is essential for PBR-VANET. Also, the application simulator will be equipped with virtual perception sensors, which made it capable of simulating the Cooperative Awareness applications.

The remainder of the paper is organized as follows. Section II describes some related works includes *Intelligent Transportation Systems, network and application simulators* and an novel work of *obstacle radio propagation extension* for the Network Simulator 3 (NS-3). Section III analyzes the functionalities should be fulfilled by the VANET application simulator and additional requirements to the improvements. Section IV gives an overview of the new framework with three parts, to provide the capabilities mentioned above. Section V and VI details the design and implementation of positioning and perception respectively. Section VII shows two use cases with the extensions, one per function. Finally, Section VIII summarizes our contributions and briefly explores directions for future work.

II. Related work

This section presents various related work in two parts: Intelligent Transportation System and Network Simulators.

A. Intelligent Transportation System

The road network is inter-connected among countries and there are few barriers and using the network, the vehicles easily cross country border. For the interoperability among the countries, cooperative ITS needs to be developed based on the same architecture, protocols and technologies. Europe has huge necessity of standardization of Cooperative ITS and European Commission (EC) published the action plan [4] followed by ITS standardization mandate [5], to promote the deployment of these systems in Europe. In US, the Institute of Electrical and Electronics Engineers (IEEE) is standardizing Wireless Access in Vehicular Environments (WAVE) architecture in IEEE 1609 family of standards [6] as well as IEEE802.11 variant for vehicular communication as IEEE802.11p [7]. Cooperative ITS and vehicular communications became essential for the cooperation of multiple entities in the road traffic (*i.e.*, vehicles, roadside infrastructure, traffic control centers) in order to achieve shared objectives (safety, efficiency, and comfort).

GeoNetworking [1] has been standardized by ETSI as a network layer protocol. It integrates several Position-Based Routing (PBR) strategies, including Greedy Forwarding (GF) [8] (also known as *GPSR*), which chooses a directly reachable node that is closest to the destination according to the GPS location obtained by the *Location Service(LS)* request action, to route packets more effectively in vehicular networks.

In the literature, the evaluation of GeoNetworking can be performed in flexible and large-scale simulated the network with low cost. However, mere simulations cannot provide realistic evaluation results for a specific implementation of GeoNetworking. In contrast, the experimental evaluation using the implementation in a field operational testbed gives real results in the deployment phase of GeoNetworking. Though in practice, it requires a heavy cost to conduct the experiments regarding time, manpower, space, and expense. To take the benefits of real field test and simulation, a realistic network simulator with geo-positioning features is necessary.

In Vehicle-to-Vehicle (V2V) architectures, the "I-am-here" messages are widely employed to notify the surrounding about the position of the vehicles. In the ITS station architecture, this type of message is named as *Cooperative Awareness Message (CAM)* [3].

CAMs are messages exchanged in the ITS network between stations to create and maintain awareness of each other and to support the cooperative performance of vehicles using the road network. A CAM contains status and attributes information of the originating ITS-S, e.g., time, position, motion state, activated systems. On reception of a CAM, the receiving ITS-S becomes aware of the presence, type, and status of the originating ITS-S.

Due to realistic reasons, such as market penetration ratio, CAM could not be required to be deployed on every vehicle. This may cause a serious problem because the ITS stations cannot be aware of vehicles do not equip with ITS infrastructures. To solve this problem, [9] have proposed an infrastructure-bases Proxy CAM under some environments. It employs roadside units (RSUs) to detect non-ITS objects and generate and broadcast CAMs in the behavior of them.

In the proposal, RSUs are assumed to be equipped with sensors in charge of object detection. On the contrary, in the experiment setup, there is only one stereo camera, because of the limitation of cost and manpower. In other words, the work needs to be evaluated in large scenarios. The only promising way to achieve this is the simulation.

B. NS-3 Direct Code Execution

NS-3 is a discrete-event network simulator, which is widely used in networking researches. NS-3 Direct Code Execution (NS3-DCE)[10] is an application simulator based on NS3. It mainly features reproducible experiments and easy debugging.

In the paper, the authors emphasized experiments reproducibility which was defined by [11]: experimentation realism, topology flexibility, easy and low-cost replication.

NS3-DCE takes a Library Operating System (LibOS) approach to making a slightly modified Linux kernel network stack running in the simulated environment. It implements a standard-compatible POSIX layer for user applications to be built onto, with no or minor modification. It makes developing and testing new protocols easier and more predictable. Also, it achieved a very good performance compared to the other works such as Mininet.

NS3-DCE have a modified Linux kernel layer and its own implementation of POSIX layer. Subsequential researchers are made easy to extend it to realize new functions or improve the performance.

C. Obstacle and Radio Fading Model in NS-3

Carpenter et al. [12] proposed an obstacle model implementation for the NS-3, in order to simulate radio shadowing in the NS-3 environment.

The radio shadowing model is separated into three layers. First, it uses a *polygon* to hold the outline of an obstacle (e.g., outer walls of a building). Then, it defines the *Topology* to handle the set of the *polygons*, which including the handling the data imported from OpenStreetMap. Finally, it uses the Computational Geometric Algorithms Library (CGAL) to calculate the radio shadowing according to the *Topology*.

In the implementation, it extended the *ns-3* in two parts: the `core` part and the `radio propagation` part. Three main classes are defined as follows:

`Obstacle` implements the `polygon` related data structures. This class is an extension of the `core` layer.

`Topology` operates on the set of `Obstacles`, which includs handling of the data imported from the OpenStreetMap. This class is also an extension of the `core` layer.

`ObstacleShadowingPropagationLossModel` extends the standard interface

`PropagationLossModel`, which calculate the path propagation loss by the obstacle information. It could be plugged into any user program with the `PropagationLossModel` compatible interface.

The paper presented a concrete and promising implementation of the radio shadowing model. However, work only provides the model for radio shadowing. Thus, we need a different model for perception calculation. Fortunately, we may re-use the infrastructures including polygons and topology to get rid of re-inventing the wheel.

III. Problem Statement and requirements

In the last section, we have detailed the missing functionalities for network and application simulators and shown some related works about obstacles. In this section, we will analyze the issues we've found in the last section and detail the desired functionalities of the system, then define some additional requirements to our proposal.

A. *Position-awareness in application*

Obviously, every ITS applications requires knowing the position of the vehicle, either directly or indirectly.

In order to bring the positioning to the applications, we need to fill the gap between Network simulator and ITS applications, by extending the application simulator. Two additional requirements should be fulfilled.

Keep standardized protocol

Standardized protocols are important because user applications usually builds according to the standard. If we can stick with the standard protocols, user applications can be simulated directly without any changes. This benefits our user (researcher) on both programming simplicity and does not require extra testing / debugging regarding the protocol changes.

In our case, Global Navigation Satellite System (GNSS) could be selected. Among the GNSSes, the most widely known and used system is the Global Positioning System (GPS). Most of the GPS devices follows the "NMEA 0183" standard. NMEA 0183 is an ASCII character and sentence based protocol running on the serial RS-232 port. It also allows simultaneous listens to a same device, a.k.a broadcasting. These features made it relatively easy to be implemented.

Experimental reproducibility

As we described above, experimental reproducibility is essential to application simulator. Experimental reproducibility requires the simulation to the positioning device fully determinative. The output of the device should be completely same between different simulations (and unrelated simulation configurations). This requirement may bring extra challenges to the design of the simulated device.

B. *Object-detection functionality in NS-3 and application*

In robotics, perception (or machine perception) means the ability a computer receives the real world data and interpret them similar humans beings. In the ITS, perception is also a fundamental requirement, since the autonomous vehicles should always aware of what happens the around them. In the most basic period, we narrow the term "machine perception" into "machine vision", which itself is based on object-detection functionality.

Nevertheless, the object-detection function is missing on both NS-3 and DCE. Thus, we need to implement it from scratch in NS-3, then provide a simulated sensor interface to the user program through DCE. Two requirements are defined in the design, as well as the most basic one: experimental reproducibility.

Scalable to large-scale scenario

Performance and scalability should be an important consideration. If we come to a factorial complexity ($O(n!)$) approach, even fastest computers cannot make the simulator scale into hundreds of vehicles. In order to conduct large-scale experiments, we should optimize the computational complexity by workaround the time-consuming part, and make sure it is well scalable.

Able to accommodate different types of sensors

In the market, various types of perception sensors based on different technologies are developed, including vision-based, lazer-based, IR-based, doppler-based, etc. Different types of sensors greatly differ on functionalities and performances. The simulator should be able to accommodate these types of sensors, to fulfill the requirements of experimented applications.

IV. Overview of Position and Perception simulation

In this section, we will discuss how to realize the position and perception in NS3-DCE and give out our proposals.

Hereby, we define *full-simulation* and *para-simulation* which are similar to the widely-used term *full-virtualization* and *para-virtualization* [13]:

Full simulation

Fully reflect the behavior of the simulatee, which means the interface to the device and the protocol are completely identical as the real device. In our case, a full-simulated device does not require any modification to the user program.

Para-simulation

A technique to provide an interface to the device, but not identical to the real underlying hardware. This approach may boost the performance by bypassing unnecessary work from the protocol and data representation. In our case, a para-simulated device requires a user-mode daemon or plugin to the user program to cooperate with.

Figure 1 is a simplified module graph of the NS-3 and DCE, with the extended modules. Three new modules are required: **GPS** and **Perception Sensors** in the DCE, and **Object Detection** in the NS-3.

V. Design and Implementation of Positioning

In the last section, we proposed a architecture considering the extension to NS-3 and DCE. This section details the design and implementation of the positioning module.

A. *Position augmentation, stablize and coordinate conversion*

In order to generate GPS signal, we need to convert the cartesian coordinate in the NS3 to the GPS coordinate (i.e., latitude and longitude). Several parameters are defined to make the simulation scalable to different GPS devices.

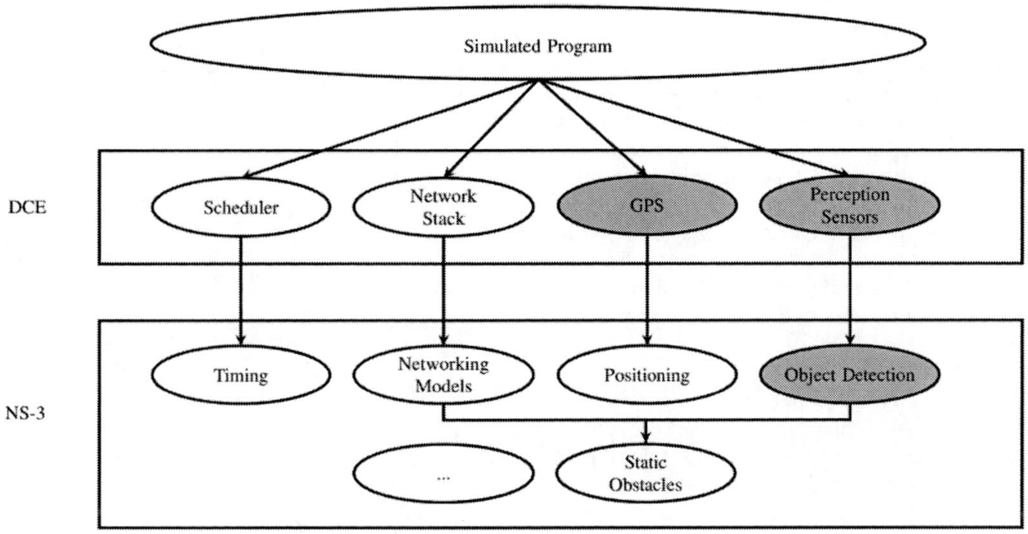

Figure 1. NS3-DCE architecture and extended modules

center The center coordinate of the experiment, in (latitude, longitude).
accuracy The accuracy of the simulated GPS device, in m.

To make the node position more realistic, we did some data augmentation to add randomness to it. For the experimental reproducibility, the randomization approach should be fully deterministic. NS-3 provides deterministic Random Number Generators (RNGs) including different distributions. In the case, we use a `NormalRandomVariable` with $\sigma = accuracy \div 3$ for the radius and a `UniformRandomVariable` with range $[-\pi, \pi)$ for the angle.

Then we did a sensor fusion to stabilize the randomized position using the Kalman Filter Algorithm. The internal variable of the filter is stored with the state of RNGs in the `DceNodeContext`, which is used for keeping the per-node information, in order to keep the experimental reproducibility among different experimental setups. Finally, the NS-3 coordinate is converted into GPS coordinate.

B. NMEA full-simulation

We take the fully-simulation approach because of the requirement of *keeping on standard format*, i.e., the NMEA format.

The data flow is shown in Figure 2. A unix character device `/dev/ttyGPS` is exposed to the user program. The device driver is defined in the POSIX layer and named `UnixGPSttyFd`. When the device is read by a program, `DCENodeContext::GPSttyRead()` is called. Then the function call another function `DCENodeContext::GetGPSPosition()` to get the "GPS position" from the mobility model in NS-3.

After the `DCENodeContext::GetGPSPosition()` got the node's real position according to the mobility model, it calculates the GPS position as defined in Section V-A. Then the function returns to the `DCENodeContext::GPSttyRead()`. Here, the function

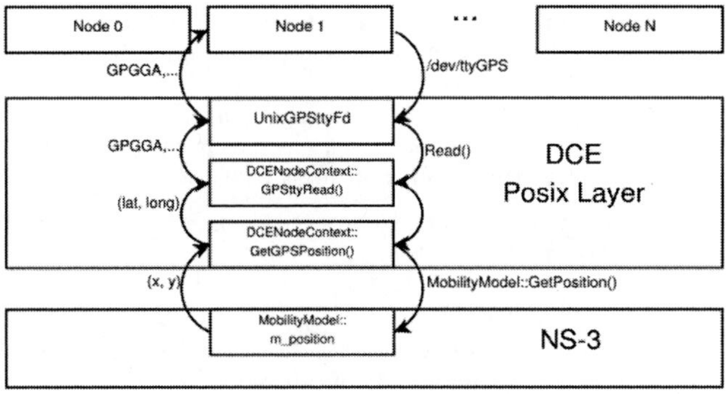

Figure 2. GPS device data flow

generate the necessary GPS flavor NMEA clauses (specifically `GPGGA` and `GPRMC`) and finally return to the caller program.

VI. Design and Implementation of Perception

This section describes an on-going work towards the perception functionality of NS-3 and DCE. According to the architecture defined in Figure 1, vehicle perception sensor simulation on the *DCE* is based on the object detection function on *NS-3*.

A. Architecture overview

In NS-3, polygons and some of its operations are supported through the `CGAL` by the obstacle extension module contributed by [12]. Additionally, the module comes with an approach to import obstacle information from the OpenStreetMap (OSM). These functions can be reused as a common infrastructure in NS-3 to represent and handle obstacles, with several new functions.

On the other side, the vehicles themselves are also polygons. We can define the outline of each type of vehicles as 2D polygons, and then they can be awared in the calculation. Other types of objects (e.g., non-its vehicles, non-motor vehicles, pedestrians) can be defined as "stub nodes", i.e., nodes do not run any protocol stacks or programs on it.

Based on the static obstacle and vehicle outline representation, the object detection simulation could be realized. When an object-detection is requested on a sensor, it counts all the vehicles (including stub ones) which are within or intersects with its sensing range. Then a line-of-sight check is performed to check each vehicle in the range is visible, or fully blocked by either static obstacles or another vehicle at the moment. Finally, for each visible vehicle, an `ID` is assigned to it, for both NS-3 representation and simulated sensor representation.

B. Modeling the sensors

Currently, different types of perception sensors are available on the market. According to [14], vehicle perception sensors can be classified into different types, including vision-based, lazer-based, IR-based, radar, doppler, ultrasonic, induction loop, magnet field, etc. To accommodate different types of perception sensors, several sensor parameters are defined:

position Coordinate of the sensors, in (x, y). Follows the position of the node, if not specified.
range Shape of the sensible region of the device, in one of the Circle_2, Circular_arc_2, Polygon_2.
min_size Minimum object size (in solid angle) which could be detected.
accuracy Accuracy of the simulated sensor, in m.
capability Mask of capabilities the sensor has (identify vehicles, classify vehicles, or just count)

For instance, several sensor products could be modeled with parameters in Table I.

C. The Para-simulation

Different from the position part, we decided to follow the para-simulation approach for the perception. The reasons are as follows:

Firstly, the vehicle perception sensors do not have a widely-used standard protocol: different products require user applications to communicate in different protocols, and also they need different routines to achieve the object-detection from the device data. In this case, extra developing will always be required for a new type of device, including our simulated ones.

Secondly, object-detection from the images or point-clouds are very time-consuming, as well as image generation, if we use a full-simulation approach. To simplify and speed-up the simulation, we bypass the *Object → Image → Object Detection* routine and directly offer the result of object detection to the user program. We model different types of the perception sensors and define several arguments to imitate the behavior and performance of each sensor.

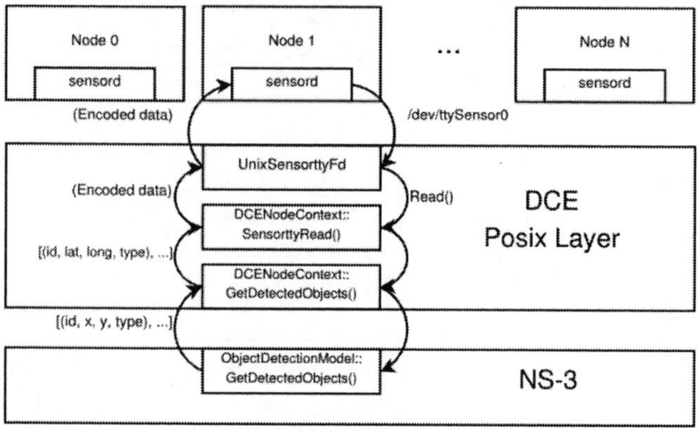

Figure 3. Perception sensor device data flow

The data flow is shown in Figure 3. Unlike the GPS approach described above, we inject a program `sensord` to the user environment, which is in charge of communication and object-detection. `Sensord` comes with an SDK with libraries to access the data, like other real sensor hardware does.

The `sensord` program communicates with the POSIX layer with the character device `/dev/ttySensor0`. The posix layer processes in a similar way as the GPS device, and finally the request comes to the NS-3 Object detection model.

The model returned the raw data which is presented as a list of *id, x, y, type*, and is subject to extend. Then the `DCENodeContext::GetDetectedObjects()` function converts the *(x,y)* coordinate into the GPS coordinate, with the help of the utility from GPS model. The function `DCENodeContext::SensorttyRead()` encode the data in json (to keep flexibility to extensions) and return to the user space daemon `sensord`. `Sensord` decodes the data and return to its subscriber (i.e., the user programs in the same node).

VII. USE CASES

Here, we show some example that our extensions could be employed. Both positioning and perception sensing are most common tasks in ITS. Our work will be useful for various kinds of scenarios and not limited to the following ones.

A. Position Based Routing

The Position-based routing algorithms are not perfect enough and needs improvements on different issues. GeoNetworking is not an exception. We presented a proposal named "Duplicated Unicast Packet Encapsulation" [15] to improve the reliability of the GeoNetworking protocol for important multi-hop messages by employing overlay networking and

TABLE I. PARAMETERS FOR DIFFERENT TYPES OF SENSORS

Sensor Type	ZMP RoboVision II	Velodyne LiDAR	Diamond Phoenix II
Type	Stereo camera	Laser Radar	Induction Loop
Position	In-vehicle / Fixed	In-vehicle / Fixed	Fixed
Range	Sector, 80m	Circle, 100m	Circle / Rectangle, 1m
Min Size	50cm, 10cm	...	10cm, 10cm
Accuracy	30cm	...	...
Capability	Identify	Identify	Classify

multipath routing. It presents an implementation based on an open-source ITS network stack, and several (real and simulated) experiments were conducted on the "Combined Realistic Evaluation Workflow" [16]. However, the vehicle movement scenarios in the simulations are "hand-crafted" and not realistic enough nor scalable enough. With the proposed work, we can conduct experiments with real street maps, to make the experiments scalable and realistic.

B. Cooperative Awareness Messages

In order to keep compatibilities with the plain old manual-driving vehicles, some transition techniques of Cooperative Awareness Service is required. Kitazato et al. [9] proposed a preliminary work call "Proxy CAM". By using perception techniques provided by sensors, road-side units (RSUs) could aware of the non-CA-ready vehicles and send CA messages (CAMs) on their behalf. In the paper, they conducted several simple and small-scale experiments using three to four nodes. Apparently, those are far not enough to evaluate the proposal in a large scenario.

A realistic simulator with sensors enabled could greatly help the evaluation. We can build flexible scenarios based on real maps of different cities, and deploy various types of position sensors on various type of objects, e.g., LiDARs and stereo visions on the ITS enabled vehicles, dopplers on traffic lights, induction detection loops under the road.

VIII. Conclusion

In this paper, we proposed methods to realize positioning and perception devices in the application simulator NS3-DCE. Both work we proposed satisfy the "experimental reproducibility" request, i.e., producing deterministic result among different simulation runs. Additionally, we attempted to model different types of perception sensor techniques and bringing adjustable parameters for flexibility to the users of the simulator.

The design and implementation of perception are not finished yet. In the near future, we will finish the following targets:

Update Location Module
We have built a prototype location module with basic function (i.e., NMEA simulation). The data augmentation is not implemented yet. We will rewrite the location module with the data augmentation and other features.

Implement 2D Perception
Currently, 2D perception is in the design period. In the future, design and implementation challenges may occur. The challenges should be overcome or workaround.

Evaluation
After the location module is updated and 2D perception module is implemented, we are able to conduct experiments on it. With the help of the modules, large-scale experiments using the real world map become possible. We will evaluate using different applications, maps of different cities, and different types of the perception sensors.

Acknowledgment

We would like to thank Xin Li of our lab, for his groundbreaking ideas of the prototype GPS mocking implementation, e.g., intercepting device handler for the GPS device. We also thank Masahiro Kitazawa, for his knowledges and ideas on the characteristics of various types of perception sensors.

This work was supported by JSPS KAKENHI Grant Number JP17H04678 and JP26730045.

References

[1] Intelligent Transport Systems (ITS); Vehicular Communications; GeoNetworking; Part 4; Sub-part 1, Std., Jul. 2014. [Online]. Available: http://www.etsi.org/deliver/etsi_en/302600_302699/3026360401/01.02.01_60/en_3026360401v010201p.pdf

[2] ISO 21217:2010 Intelligent transport systems – Communications access for land mobiles (CALM) – Architecture, ISO CALM TC204 Std., April 2010.

[3] Intelligent Transport Systems (ITS); Vehicular Communications; Basic Set of Applications; Part 2, Std., Sep. 2014. [Online]. Available: http://www.etsi.org/deliver/etsi_en/302600_302699/30263702/01.03.01_30/en_30263702v010301v.pdf

[4] Action plan for the deployment of Intelligent Transport Systems in Europe, European Commission Std., December 2008, cOM(2008) 886 final.

[5] Standardisation mandate addressed to CEN, CENELEC and ETSI in the field of information and communication technologies to support the interoperability of co-operative systems for intelligent transport in the european community, EUROPEAN COMMISSION Std., October 2009.

[6] IEEE 1609.0 Draft Standard for Wireless Access in Vehicular Environments (WAVE) - Architecture, IEEE Std., April 2010.

[7] IEEE Standard for Information technology - Telecommunications and information exchange between systems - Local and metropolitan area networks - Specific requirement, Part 11: Wireless LAN Medium Access Control (MAC) and Physical Layer (PHY) Specifications, IEEE Computer Society Std., July 2010, iEEE Std 802.11p-2010.

[8] B. Karp and H. T. Kung, "Gpsr: Greedy perimeter stateless routing for wireless networks," in 6th Annual International Conference on Mobile Computing and Networking, MobiCom 2000, August 6.-11., 2000, Boston, Massachusetts, USA. ACM / IEEE, August 2000, pp. 243–254.

[9] T. Kitazato, M. Tsukada, H. Ochiai, and H. Esaki, "Proxy cooperative awareness message: an infrastructure-assisted v2v messaging," in Mobile Computing and Ubiquitous Networking (ICMU), 2016 Ninth International Conference on. IEEE, 2016, pp. 1–6.

[10] H. Tazaki, F. Uarbani, E. Mancini, M. Lacage, D. Camara, T. Turletti, and W. Dabbous, "Direct code execution: Revisiting library os architecture for reproducible network experiments," in Proceedings of the ninth ACM conference on Emerging networking experiments and technologies. ACM, 2013, pp. 217–228.

[11] N. Handigol, B. Heller, V. Jeyakumar, B. Lantz, and N. McKeown, "Reproducible network experiments using container-based emulation," in Proceedings of the 8th international conference on Emerging networking experiments and technologies. ACM, 2012, pp. 253–264.

[12] S. E. Carpenter and M. L. Sichitiu, "An obstacle model implementation for evaluating radio shadowing with ns-3," in Proceedings of the 2015 Workshop on ns-3. ACM, 2015, pp. 17–24.

[13] A. Whitaker, M. Shaw, S. D. Gribble et al., "Denali: Lightweight virtual machines for distributed and networked applications," Technical Report 02-02-01, University of Washington, Tech. Rep., 2002.

[14] T. V. Mathew, "Automated traffic measurement," Trac Engineering And Management, IIT Bombay April, 2012. [Online]. Available: https://www.civil.iitb.ac.in/tvm/1111_nptel/524_AutoMer/plain.pdf

[15] Y. Tao, X. Li, M. Tsukada, and H. Esaki, "Dupe: Duplicated unicast packet encapsulation in position-based routing vanet," in 2016 9th IFIP Wireless and Mobile Networking Conference (WMNC), July 2016, pp. 123–130.

[16] Y. Tao, M. Tsukada, X. Li, M. Kakiuchi, and H. Esaki, "Reproducing and extending real testbed evaluation of geonetworking implementation in simulated networks," in The 10th International Conference on Future Internet. ACM, 2015, pp. 27–34.

Application of Deep Learning to Route Odometry Estimation from LiDAR Data

Miguel Clavijo, Francisco Serradilla, José E. Naranjo, Felipe Jiménez and Alberto Díaz
University Institute for Automobile Research (INSIA)
Universidad Politécnica de Madrid (UPM)
Madrid, Spain
e-mail: miguel.clavijo@upm.es

Abstract— **The Deep Learning techniques are a powerful tool to support the development of all sorts of information classification or processing techniques within the area of intelligent vehicles, since they are able to emulate the performance of the human brain when learning from experience. Specifically, the technique of Convolutional Neural Networks (CNN) has been successfully used in applications for classification and localization of pedestrians and obstacles on the road. However, CNN allow not only classification and pattern learning, but can be used for regression or modeling, like other kind of classical neural networks. The fundamental difference of both applications is that, while in classification the values of the network output are usually discrete, in regression or modeling applications the network can generate a continuous output with real numbers, allowing it to emulate the output of any type of system that is presented in the training set, with all its associated advantages, such as generalization and correct characterization of situations that have not learned explicitly. This paper presents an application of CNN for modeling in Intelligent Vehicles field, whose objective is to calculate the navigation parameters of a vehicle from the information supplied by a 3D LiDAR mounted on a vehicle that circulates in urban areas. Specifically, the developed CNN is able to calculate the speed and heading of a vehicle circulating in real time from the distance data supplied by the LiDAR sensor. The results show that the network is able to learn to calculate the speed and the yaw rate from the identification of the characteristic points of the environment, providing data that can be used to support the navigation of the vehicles.**

Keywords - deep learning; autonomous vehicle; odometry; LiDAR.

I. INTRODUCTION

Intelligent vehicles are characterized by equipping a large number of sensors and computer and communication systems, capable of providing all the information required for some advanced driving assistance systems, such as cooperative systems, specifically on the field of autonomous vehicles navigation. This large number of sensors allows monitoring the driving environment with high precision, even beyond the visual horizon due to the communications systems and, at the same time, supporting a safe navigation. However, this instrumentation has two fundamental problems. On the one hand, the high number of sensors (Computer vision, 3D LiDAR, Ultrasounds, Radar, Gyro, Compass, GPS, etc.) requires a large investment in these vehicles, increasing production costs. On the other hand, many of these sensors provide redundant information, which may or may not be used by the system, which in many cases underuse this data.

In this way, one of the most widely used sensors in the environment recognition of intelligent vehicles is the Laser Scanner 3D or LiDAR. This type of sensors usually provide an array of points with the distances from the sensor to the different elements of the environment in a range of 360°. Generally, this accurate information of the driving environment is used for the detection of pedestrians, vehicles or other obstacles on the road. However, since this precise information about the environment is available, it is possible to use it for other applications, such as navigation support [1] or visual odometry.

Visual odometry has been one of the last research fields to take part into the autonomous navigation applications. In general terms, visual odometry techniques tackle the SLAM (Simultaneous Localization And Mapping) problem, estimating the vehicle ego-motion and locating it in an unknown environment by using perception sensors as main source of information. Both estimating an accurate motion and tracking the vehicle route are two of the most difficult tasks in robotic and therefore in autonomous vehicle development. Solving the SLAM problem allows to perform critical tasks, such as the autonomous navigation where GPS signal can be lost or driving through complex areas, among others. In recent years, different visual odometry techniques have been developed using computer vision, stereo vision [2], LiDAR [3] or a sensory fusion between them [4][5]. Each of these algorithms, extract specific features of the environment such as flat surfaces, vertical corners, sharp angles, etc., from the data supplied by the sensor in each case and followed by a matching process between frames. Furthermore, in some cases it is necessary to make use of external motion sensors (e.g., IMUs, GPS/INS) in order to decrease the error. Another solution employed is the "loop closure" method when the incremental errors over time produce some drift. This implies that it does not work in real-time.

On the other hand, one of the most promising techniques for application, in multiple domains in general and in intelligent vehicles in particular, is Deep Learning. Deep Learning comprises a set of intelligent and bioinspired techniques based on neural networks with multiple hidden layers (usually more than 3). Convolutional Networks, autocorrelators, deep belief networks and Long Short-Term Memory (LSTM) are the four basic neural network techniques that establish the Deep Learning framework. Convolutional layers in a network are specialized in image processing and are those that learn the convolutions to perform on the input data (image pixels) to filter it or obtain relevant features. Specifically, the convolution processes an input image to obtain its relevant features. Adding several Full Connected traditional layers to the convolutional layers, the resulting network is able to identify patterns in the images, classifying them as belonging or not to a particular class or standard (e.g., pedestrian or car), and are widely used as classifiers for camera-based systems in the field of intelligent vehicles.

Due to all these characteristics, CNN is starting to be used in autonomous vehicles applications. As mentioned, this type of networks is mainly used in classification and pattern learning when the number of outputs is discrete. Because of this, several works have implemented CNN based system to identify obstacles in autonomous navigation [6]. On the other hand, CNN are not only being used for classification, but also for estimating the vehicle ego-motion. In this way, according to [7], using stereo vision and extracting features, is possible to estimate vehicle velocity and direction, even though it calculates only discretized values. In turn, in [8] LiDAR information is used to estimate odometry using regression, which entails, continuous values as output. However, in that work, voxel grids are used to extract generic features.

In this paper, a novel application of convolutional neural networks within the field of intelligent vehicles is presented. Thus, a CNN based system is proposed that is able to calculate the speed and yaw rate of a vehicle that circulates in urban areas at speeds up to 50 km/h. Specifically, it is proposed to use this type of networks using as input the information of the point cloud provided by a 3D LiDARIn addition, using this type of Deep Learning technique, allows a better understanding of the potential that CNNs have. Furthermore, this special application for autonomous vehicles is crucial, due to the high computational cost of the system when using a great number of data and classic programming methods. In this way, the CNN used is able to supply a precise output of the navigation parameters of the ego-vehicle and reduce drastically the computational cost once the network model is trained. Pre-training or environment features extraction is not previously made. This system has been implemented, trained and tested in the facilities and with the vehicles of the University Institute of Automobile Research (INSIA) of the Technical University of Madrid, obtaining results comparable to the data supplied by high performance speedometer, GPS and gyroscope.

The paper is structured as follows, in Section II, the architecture of the CNN used is described as well as how the training process has been tackled. In Section III, the procedure for transforming the raw data to the image-data for the network input is adressed. The results of speed and yaw rate obteined in the test dataset are shown in Section IV. Finally, in Section V, the conclusion and further works are dicussed.

II. CONVOLUTIONAL NETWORK ARCHITECTURE, PARAMETERS AND TRAINING FRAMEWORK

A. *Architecture description*

A Convolutional Neural Network is a feed forward neural network mainly composed of convolutional layers. Those layers are the matrices shown in (1)

$$H \times W \times C \qquad (1)$$

Where H and W are spatial dimensions and C the number of channels (or channel dimension).

The architecture presented here is the result of a number of trial-and-error steps with different network sizes (both number of layers and convolutional networks dimensions), learning methods and parameters. The final architecture is summarized in Figure 1. The network is made up of six layers – four convolutional layers and two fully connected.

The input to the network is presented as 8×300×2 images, where the two channels corresponds to the image created from the points data captured by the LiDAR at "*t*" and "*t-1*". Then, features are extracted through the convolutional layers, adjusting the resulting features to the final two values (speed and yaw rate) using two fully connected layers.

Each convolutional layer is composed of three or four operations or sub-layers:

- The convolution operation itself, where, given an image with the pattern defined in (1), a set of filters ($N \times M \times C$ matrices where N≤H and M≤W)

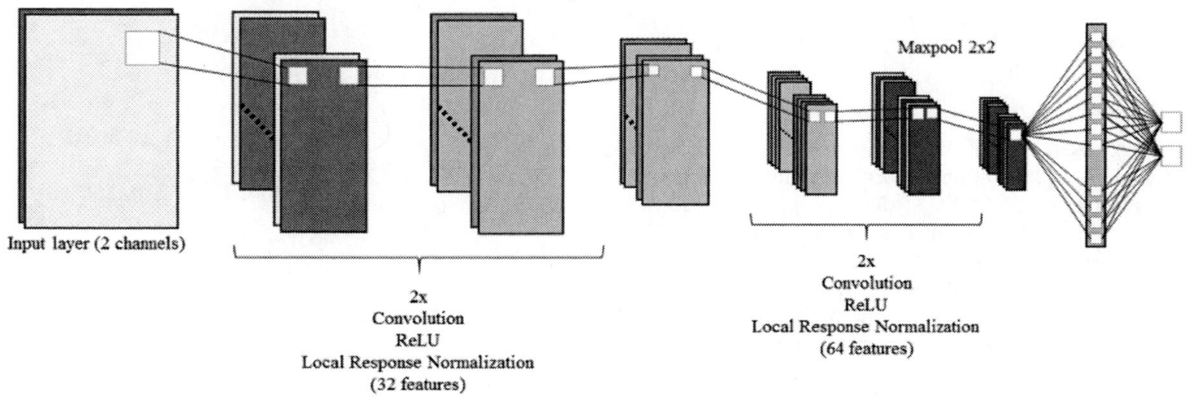

Figure 1. Convolutional network architecture proposed.

traverse the image with a step of 1 extracting 1 value for each sub-image.

- A Rectified Linear Units (ReLUs) [9] network, with neurons whose activation function is:

$$f(x) = \max(0, x) \quad (2)$$

Its purpose is to introduce non-linearity to the net. Other functions like (3) and (4) can also be used but they are much slower in terms of training time than ReLUs [9].

$$f(x) = \tanh(x) \quad (3)$$

$$f(x) = \frac{1}{1+e^{-x}} \quad (4)$$

- Local response normalization operation [10], whose aim is to increase the differences (i.e., improve the contrast) between adjacent pixels.

- Maxpool operation, is a non-linear subsampling method where a filter and a step are defined (usually a N×N filter with a step of N), replacing each subset of that window size by its maximum value.

The first and second convolution layers have a dimension of 3×15×32. The maxpool operator is applied at the end of the second layer with a 2×2 dimension with a step of 2, so the output is an image of dimensions 4×150×32, that is, 32 features. Third and fourth convolution layers have a dimension of 2×5×64. As with the second layer, a maxpool operator of same dimensions and step size is applied at the end of the fourth layer so the output for this layer has the dimensions 2×75×64, then 64 features.

The fully connected layers are traditional feed-forward neural networks where all the outputs of one layer are connected to all the inputs of the next one. In this case, the first fully connected layer transforms the 2×75×64 outputs of the last convolutional layer to a vector of 512 values. This outputs are then transformed into the last 2 values by second fully connected network.

Finally, a learning-with-dropout scheme has been used, with a dropout rate of 0.9 (10% of the neurons are removed each iteration). The dropout technique [11] is a way to avoid overfitting by randomly removing some of the neurons in the network each iteration. This way of learning makes the neurons not to learn by memory, but sharing knowledge among several instead.

B. Learning operation

The proposed model was trained using a stochastic gradient-based optimization algorithm called Adam [12] with a learning rate of $1e10^{-5}$ throughout 3,000 iterations. It has been tested with different learning rates: $1e10^{-3}$, $1e10^{-4}$ and $1e10^{-5}$. With a learning rate of $1e10^{-3}$ and $1e10^{-4}$, the convergence behavior obtained suggested that it was too high for the topology of the particular problem. Therefore, it has been chosen to use this value of learning rate.

The network was developed in the Python programming language with the help of the TensorFlow library [13] for the modelling and parallel training setups.

For the selected training dataset, this convolutional network configuration took 55 hours on a computer with GNU/Linux, a Xeon E3-1200 (family) v3/4th Gen Core Processor and a NVIDIA GTX 980Ti with 6GB and 2816 CUDA cores. As mentioned above, the execution time once the network is trained, allows the performance in real time.

III. DATASET AND TESTS

The architecture of the described network uses raw data obtained by the 3D LiDAR as input. This makes possible that the implementation of the trained network in the on-board computer be simpler and not dependent on extra sensors. As outputs, this network estimates the vehicle speed and its yaw rate.

In order to create the dataset of the network, a pre-treatment of the LiDAR data was performed. As ground truth, it has been used the speed data available from a speedometer and the yaw angle acquired by a high accurate Gyro, placed on the vehicle.

For a neural network training, it is necessary that the number of inputs and outputs is the same for all the dataset. Furthermore, when using convolution networks, a type of images has been developed from where the features are extracted.

The creation of these data-images requires several processing steps of the raw data of the laser scanner for each frame:

- First, the laser field of view (FOV) range is defined. Vertically, the range of vision is set between 3° and -11°, dividing the laser points in 8 rows. The horizontal FOV comprises two 150° areas located on both sides of the vehicle and placed centrally on its transverse axis (Figure 2). This range of vision was chosen given that it contains the area in which the environment perception generates the better amount of information for the estimation of odometry.

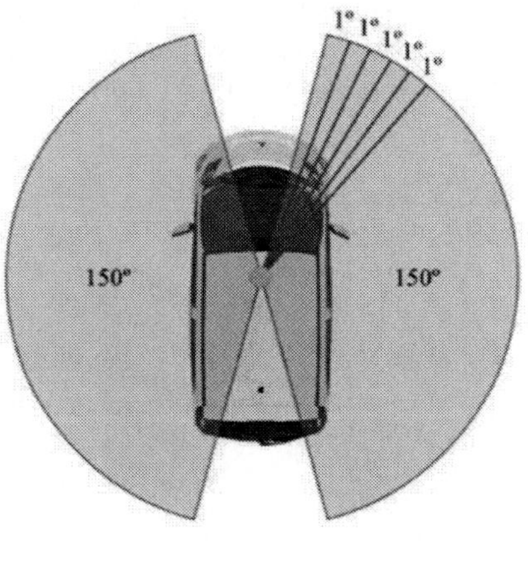

Figure 2. Horizontal FOV and its discretization.

- The LiDAR sensor used does not guarantee the generation of the same number of point in each revolution. Thus, in order to obtain the same amount of the data in each frame, the horizontal FOV is discretized with an amplitude of 1° (Figure 2). The distance of the LiDAR-points is acquired and these points are classified according to its horizontal angle and the channel through they were obtained. For this reason, 300 values are acquired in each of the 8 channel defined previously. This will be the data image resolution. In the case that more than one point exists for the same "pixel", the one presenting the lowest distance value remains.
- The disadvantage of LiDAR versus computer vision technology concerns the data dispersion when the distance is greater. Due to this, obtaining data that does not satisfy the conditions of the "pixel" is a possibility. Therefore, when a "pixel" with a null value is acquired in the image, a linear interpolation between the near valid values is done. In addition, as mentioned in other works [14], points density per length unit is inversely proportional to the distance to the sensor. Thus, in order to compensate the values distribution, a logarithmic filtering is applied in each pixel-value, followed by a normalization between 0 and 1, according to the distance. As a result, the histogram after this calculation is widely distributed throughout the normalized distance range (Figure 3).

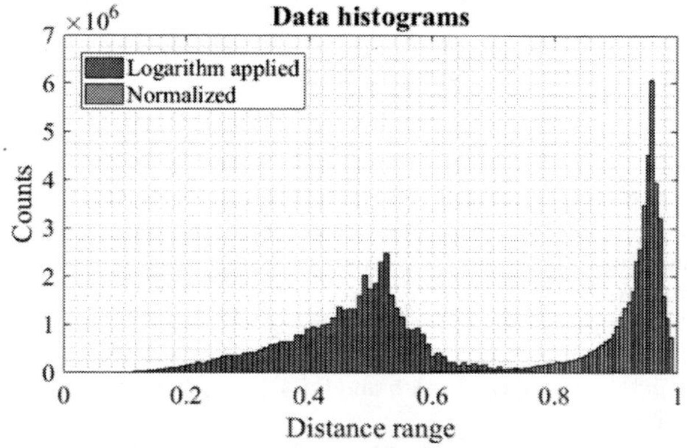

Figure 3. Data with and without logarithm applied.

As a result, in Figure 3 the original normalized data is shown in red. This value distribution is concentrated in a narrow range (between 0.8 and 1). In contrast, the data distribution after apply the logarithm to widen along the entire range is represented in blue.

- Finally, two data images are created, one for each channel of the network (parameter C in (1)) (Figure 4). The first corresponds to time "*t*" and the second is the image defined for time "*t-1*". The data acquisition is at 10 Hz, so the time transition is 0.1 s. As outputs, the average speed and the yaw rate during the transition time are included.

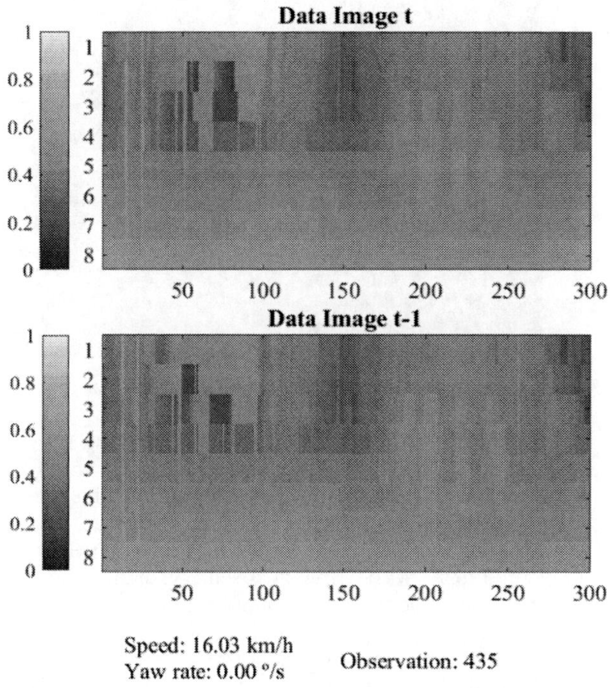

Figure 4. Example of data-images in a specific observation.

Figure 4 represents one of the images create for the CNN. The horizontal range of vision discretization results in 300 divisions, while the data is adquised by 8 channels corresponding to the vertical range of vision, defined between -11° and 3°. Therefore, the image size is 8×300 pixels. Each of those pixels is the distance data of the LiDAR points after the logarithm has been applied and normalized.

Neural network training requires a large amount of observations. This is why create a dataset for Deep Learning applications take a considerable amount of time. As a solution for this problem, a "mirror function" was implemented. This function doubles the number of observations used for training, obtaining twice the number of valid samples and taking into consideration new situations in the environment. This is possible because the "mirror function" calculates the inverse of data-images and their outputs, simulating a data acquision performed in a non-real scenario whose trajectory is the inverse of the real-scenario.

IV. RESULTS AND DISCUSSION

Several data collections were performed and the correct behavior for the network training was proven.

In this case, a route at Campus Sur - UPM has been selected, which corresponds to a one-way street with 2 lanes and cars parked on both sides of the road (Figure 5).

Figure 5. Urban area selected for the CNN training set.

Specifically, a data collection for the training set during one hour was made, the equivalent to 25.6 km in an urban environment. During this data collection, the urban scenario had dynamic traffic, tight bends, diaphanous areas, buildings, etc. In addition, the driving mode was different throughout the entire test, with varying speeds, changing lanes and making stops. The training data collection consists of 34,530 observations, which gives rise to a total of 69,060 samples when applying the "mirror function" mentioned above.

Regarding the dataset test for the network, a different route was chosen, where one part corresponded to the same zone as the training set and the other part was an unknow area. This dataset test comprises 2.85 km or 4,756 observations.

Once the network training was finished, the CNN model was applied to the dataset test. The results obtained are depicted in Figure 6 and Figure 7 for the vehicle speed and its yaw rate, respectively.

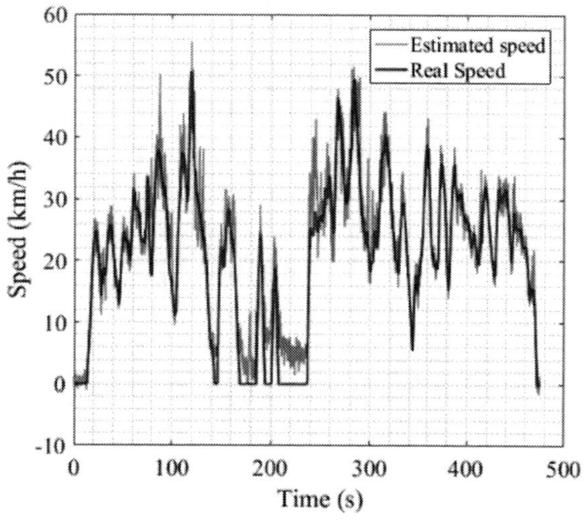

Figure 6. Speed estimation in test dataset.

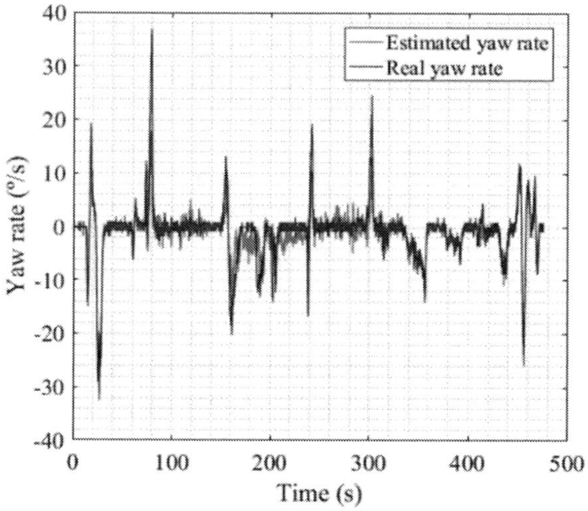

Figure 7. Yaw rate estimation in test dataset.

It can be considered that the estimation of the speed is accurate, mainly focusing on the accelerations and decelerations, where the value estimated by the network and the real value have the same outline, obtaining a Root Mean Square (RMS) error after 3,000 epoch of 3.61 km/h.

Concerning the yaw rate, the estimation is quite precise, being better adjusted to the actual data when there are concatenated path changes. An RMS of 2.8 °/s is obtained of the whole test after 3,000 epoch.

Both Figure 6 and Figure 7, illustrate that between the time instants 150 s and 250 s the estimation of the values and the real data differs somewhat more. This is largely due to the fact that it corresponds to the unknown area in which it had not previously circulated and which did not correspond of the training dataset, as previously described. On the other hand, the network was able to generalize unknown surroundings and the performance in this specific situation is reasonably successful.

It must be mentioned that the RMS obtained are in the dataset test and that it is different from the learning RMS, which has been 0.92 km/h and 0.94 °/s for the whole training set.

It can be assumed that the network, with a short training set, is capable of learning and generalizing. However, the CNN should be taught with training sets that include other road types to be able to operate in all cases. Not a great number of areas, but situations that serve as models.

V. CONCLUSION

In this paper, an application of Deep Learning techniques to estimate the visual odometry of an autonomous vehicle has been presented. Specifically, the information provided by a 3D LiDAR has been used as input of a Convolutional Neural Network with a novel architecture that is able to estimate the values of speed and yaw angle. This system has been implemented, trained and tested in the facilities of the

University Institute for Automobile Research, using its Campus as testbed area. The results of the implementation and commissioning of this system is that the CNN is able to successfully estimate the output values, generalizing correctly when in situations that have not been previously learnt by the network. Two conclusions have been achieved of the work presented in this paper: on one hand, a novel application of CNN for regression in visual odometry has been developed. On the other hand, the results of this system show that the network need typical road areas to learn how to estimate the navigation parameters and, from these data, is able to generalize the navigation of different areas not learnt.

In future works, the main aim will be improving the output precision, collecting data of specific environments and traffic situations. Also, changes in the network architecture here described, are providing useful information about the contribution of each layer. For that reason, adding additional convolutional layers may improve the overall performance.

ACKNOWLEDGMENT

This work has received the support of the European Project AUTOCITS (INEA/CEF/TRAN/M2015/1143746. Action No: 2015-EU-TM-0243-S), MINECO CAV (TRA2016-78886-C3-3-R), and Excellence Network SEGVAUTO-TRIES. The authors would also like to especially thank Óscar Gómez, who as member of the team, was responsible for setting up the test equipment and helped carrying out necessary tests to the development of this work.

REFERENCES

[1] M. Clavijo, F. Jiménez, J. E. Naranjo, and Ó. Gómez, "Supervision system for reversible lanes based on autonomous vehicles," in *11nd European Congress and Exhibition on Intelligent Transport Systems and Services*, 2016, no. June, pp. 6–9.

[2] I. Cvisic and I. Petrovic, "Stereo odometry based on careful feature selection and tracking," in *2015 European Conference on Mobile Robots (ECMR)*, 2015, pp. 1–6.

[3] J. Zhang and S. Singh, "Low-drift and real-time lidar odometry and mapping," *Auton. Robots*, no. October 2014, 2016.

[4] F. Mutz, L. P. Veronese, T. Oliveira-Santos, E. de Aguiar, F. A. Auat Cheein, and A. Ferreira De Souza, "Large-scale mapping in complex field scenarios using an autonomous car," *Expert Syst. Appl.*, vol. 46, pp. 439–462, 2016.

[5] A. Tamjidi and C. Ye, "6-DOF Pose Estimation of an Autonomous Car by Visual Feature Correspondence and Tracking," *Int. J. Intell. Control Syst.*, vol. 17, no. 3, pp. 94–101, 2012.

[6] P. Christiansen, L. N. Nielsen, K. A. Steen, R. N. Jorgensen, and H. Karstoft, "DeepAnomaly: Combining background subtraction and deep learning for detecting obstacles and anomalies in an agricultural field," *Sensors*, vol. 16, no. 11, p. 1904, Nov. 2016.

[7] K. Konda and R. Memisevic, "Learning Visual Odometry with a Convolutional Network," in *International Conference on Computer Vision Theory and Applications*, 2015, pp. 486–490.

[8] A. Nicolai, R. Skeele, C. Eriksen, and G. A. Hollinger, "Deep Learning for Laser Based Odometry Estimation," in *Science and Systems Conf. Workshop on Limits and Potentials of Deep Learning in Robotics*, 2016.

[9] V. Nair and G. E. Hinton, "Rectified Linear Units Improve Restricted Boltzmann Machines," in *Proceedings of the 27th International Conference on Machine Learning*, 2010, no. 3, pp. 807–814.

[10] A. Krizhevsky, I. Sutskever, and G. E. Hinton, "ImageNet Classification with Deep Convolutional Neural Networks," *Adv. Neural Inf. Process. Syst.*, pp. 1–9, 2012.

[11] N. Srivastava, G. Hinton, A. Krizhevsky, I. Sutskever, and R. Salakhutdinov, "Dropout: A Simple Way to Prevent Neural Networks from Overfitting," *J. Mach. Learn. Res.*, vol. 15, pp. 1929–1958, 2014.

[12] D. Kingma and J. Ba, "Adam: A Method for Stochastic Optimization," in *International Conference on Learning Representations*, 2014, pp. 1–13.

[13] M. Abadi *et al.*, "TensorFlow: A System for Large-Scale Machine Learning TensorFlow: A system for large-scale machine learning," in *12th USENIX Symposium on Operating Systems Design and Implementation (OSDI '16)*, 2016, pp. 265–284.

[14] F. Jiménez, M. Clavijo, J. E. Naranjo, and Ó. Gómez, "Improving the Lane Reference Detection for Autonomous Road Vehicle Control," *J. Sensors*, vol. 2016, p. 13, 2016.

Development of LTE based Railway Wireless Communication Systems: Preliminary Experimental Test Results

Woong Cho*, Bumgon Kim**, Hanbyeog Cho***
*Department of Computer System Engineering, Jungwon University, Goesan-gun, Goesan-eup, South Korea
** Technology Research Department, Korea Railroad Corp., Daejeon, South Korea
***Electronics and Telecommunications Research Institute, Daejeon, South Korea
E-mails: *wcho@jwu.ac.kr, ** marky3552@korail.com, ***hbcho@etri.re.kr

Abstract— **In this paper, we examine the experimental test results of LTE based railway wireless communication systems. First, the overall system architecture is introduced for railway wireless communication systems. Then, the several performance measurement results, which include Reference Signal Received Power (RSRP), communication coverage, call setup time, handover success rate, data delay time, and successive packet loss rate are introduced. Based on the measurement results, the compatibility of LTE based communication systems in railway environments can be verified.**

Keywords-railway wireless communications; LTE based system; experimental test;performance.

I. INTRODUCTION

In South Korea, several wireless communication schemes, i.e., Very High Frequency (VHF), Trunked Radio System-ASTRO (TRS-ASTRO), and TRS-TETRA, are applied in railway communication systems depending on the region. Therefore, several mobile terminals are equipped with one train to support various communication schemes which may not guarantee seamless communication environments. Furthermore, the maximum data transmission requirements are different depending on communication scheme which makes it difficult to transmit a certain data. To overcome these limitations and support seamless communication environments and deliver high data rate transmission, Long Term Evolution (LTE) based wireless communication systems have been developed. Recently, LTE scheme is considered as next generation of railway communication scheme [1][2]. Some performance results of LTE based railway communication systems are introduced in [3][4]. However, experimental measurement results adopting LTE based communication system in practical railway environments are not thoroughly introduced. In this paper, we introduce the experimental test results of LTE based railway wireless communication systems.

II. SYSTEM ARCHITECTURE

In this section, we consider system architecture. Fig. 1 represents the overall system architecture. In general, the system consists of Evolved Packet Core (EPC), Digital Unit (DU) and Remote Radio Unit (RRU).

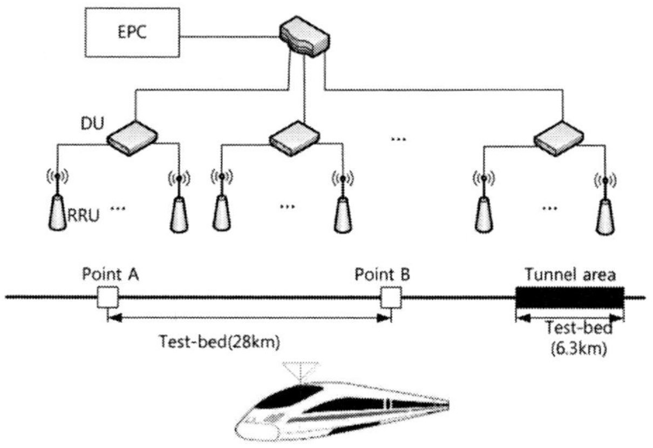

Figure 1. The overall system architecture.

EPC follows the 3GPP LTE standard structure. However, different from the general LTE networks, EPC and DU have dual system for emergency scenarios. RRU has also a dual system, which enables to transmit signal proper when the adjacent RRU is inactive status. Since railway network is directly related to the safety of passengers, the main part of system consists of dual systems. The train is equipped with mobile terminal and on board unit for communication. With this architecture, the performance of wireless communication is measured in two scenarios: one is open area and the other is tunnel area. Then, the Reference Signal Received Power (RSRP), communication coverage, call setup time, handover success rate, data delay time, and successive packet loss rate are measured. The average moving speed of train is 250km/h. The frequency band for uplink and downlink is 718~728MHz and 773~783MHz, respectively. Therefore, each link has 10MHz of bandwidth.

III. MEASUREMENT RESULTS

For experimental measurement, we transmit signal approximately with 46dBm at the end of User Equipment (UE) based on 3GPP standard in [5].

A. *RSRP and communication coverage*

The reference RSRP value is measured based on 3GPP standard, and the reference criterion is greater than or equal -110dBm. We regard RSRP as "success" if the RSRP satisfies

85% of reference value, i.e., -110dBm. Table I and Table II represent measurement results of RSRP. As we can see, if the distance between RRU and RRU is greater than 3Km, the RSRP is not sufficient.

TABLE I. RSRP MEASUREMENT IN OPEN AREA

Distance between RRU and RRU	Average measured value	Satisfaction rate
1km	-76.10dBm	100%
2km	-88.26dBm	98.8%
3km	-97.18dBm	81.9%
4km	-97.60dBm	77.7%

TABLE II. RSRP MEASUREMENT IN TUNNEL AREA

Distance between RRU and RRU	Average measured value	Satisfaction rate
0.5km	-72.58dBm	100%
1km	-79.05dBm	100%
1.5km	-84.31dBm	100%
2km	-86.61dBm	100%

B. *Call setup time*

For call setup time we measured call setup time for emergency call, Push to Talk (PTT) call setup time, and group call setup time. We also measured call success rate for each case. Table III shows reference value and its corresponding measured value. The table indicates that all cases represent successful results

TABLE III. CALL SETUP TIME

Test Condition	Reference value	Measured value
Emergency call setup time	Less than 2 sec	0.109sec
PTT call setup time	Less than 2.5sec	0.108sec
Group call setup time	Less than 2.5sec	0.151sec
Emergency call succe rate	Greater than 99%	100%
PTT call success rate	Greater than 99%	100%
Group call success rate	Greater than 99%	100%

C. *Handover success rate*

We measured handover success rate with the same condition of RSRU measurement condition. The criterion for successful handover rate is greater than 99%. Tables IV and V show that all cases fulfill the handover success rate.

D. *Data delay time*

Data delay time is measured to test the buffering time which is caused by several signal flow steps. The successful delay time is less than 600ms, where the measured average delay time is 28ms. Therefore, the measured data satisfies the data delay time criterion.

TABLE IV. HANDOVER SUCCESS RATE IN OPEN AREA

Distance between RRU and RRU	# of trial	# of success	Success rate
1km	6	6	100%
2km	6	6	100%
3km	6	6	100%
4km	6	6	100%

TABLE V. HANDOVER SUCCESS RATE IN TUNNEL AREA

Distance between RRU and RRU	# of trial	# of success	Success rate
0.5km	2	2	100%
1km	2	2	100%
1.5km	2	2	100%
2km	2	2	100%

E. *Successive packe loss rate*

Successive packet loss rate measures the loss rate when the long data packet rate (6000bytes) signal is transmitted. The measurement results show that there is no loss although long packet is transmitted.

IV. CONCLUSIONS

In this paper, we discussed the measurement results of LTE based railway wireless communication systems. The measurements are carried out by considering various performance parameters with practical implementation. The results reveal that the LTE based system can support reliable communication link in railway environments.

ACKNOWLEDGMENT

This research was supported by a grant (17-RTRP-B089546-04) from the Railway Technology Research Project funded by the Ministry of Land, Infrastructure and Transport (MOLIT) of the Korean government and by the Korea Agency for Infrastructure Technology Advancement (KAIA).

REFERENCES

[1] I. M. Liem and V. B. Mendiratta, "Mission critical communication networks for railways", Bell Labs Technical Journal, vol. 16, no. 3, pp. 29-46, 2011.

[2] J. Calle-Sanchez, M. Molina-Garcia, J. I. Alonso, and A. Fernandez-Duran, "Long Term Evolution in high speed railway environments: feasibility and challenges", Bell Labs Technical Journal, vol. 18, no. 2, pp. 237-253, 2013.

[3] J. Calle-Sanchez, E. Martinez-de-Rioja, M. Molina-Garcia, and J. I. Alonso, "Performance of LTE mobile relay node usage for uplink access in high speed railway scenarios", in IEEE Vehicular Technology Conf. Spring, pp. 1-5, Glasgow, Scotland, May 11-14, 2015.

[4] A. Sniady and J. Soler, "LTE for railways: impact on performance of ETCS railway signaling," IEEE Vehicular Technology Magzaine, vol. 9, no. 2, pp. 69-77, 2014.

[5] 3GPP TS36.101 E-UTRA; User Equipment (UE) radio transmission and reeption.

RSU Placement Method Considering Road Elements for Information Dissemination

Kazuyoshi Gomi
Keio University
3-14-1 Hiyoshi Kohoku-ku
Yokohama-shi Kanagawa Japan
Email:gomi@mos.ics.keio.ac.jp

Yusuke Okabe
Keio University
3-14-1 Hiyoshi Kohoku-ku
Yokohama-shi Kanagawa Japan
Email:okabe@mos.ics.keio.ac.jp

Hiroshi Shigeno
Keio University
3-14-1 Hiyoshi Kohoku-ku
Yokohama-shi Kanagawa Japan
Email:shigeno@mos.ics.keio.ac.jp

***Abstract*—In Intelligent Transportation System (ITS), communication with RoadSide Unit (RSU) is expected that weak connectivity of vehicle-to-vehicle communication can be improved since the power average of RSU is large. However, to deploy and maintain RSUs is costly. Therefore, it is necessary to effectively place RSUs within limited cost. Many opportunities to communicate with vehicles are at intersections with a lot of traffic volume. However, it is necessary to consider not only the traffic volume but also a connection relation of the road network because buildings prevent radio waves. In this paper, we propose an RSU placement method considering road elements that affect radio wave spreading. This method consists of two actions: calculation of RSU placement priority with considering road elements affecting radio wave spreading and operation of updating RSU placement priority. As a result of simulation, our proposal is particularly effective in a scenario considering information relay because the communication performance of this method was higher than or equivalent to that of other RSU placements. Further, it is found that our proposal was possible to suppress redundant RSU placement by updating operation of RSU placement priority in a scenario which is not considering information relay.**

Keywords*–*VANET; RSU; Road elements; Priority

I. INTRODUCTION

Recently, research on Intelligent Transportation Systems (ITS) that improves the traffic safety and traffic efficiency by making vehicles communicate with each other has been actively conducted. In ITS, vehicle communication is classified into vehicle-to-vehicle and road-to-vehicle communication. Vehicle-to-vehicle communication can be extended ad hoc networks easily because ad hoc communication is possible if only vehicles are equipped with in-vehicle devices dedicated to communication individually. In particular, ad hoc networks composed of vehicles are called Vehicular Ad-hoc NETworks (VANETs), and various applications using this VANET are being studied in the field of ITS.

In ITS, road-to-vehicle communication is expected to have high connectivity. Propagation radio waves are blocked by buildings in the city and connection of VANET is unstable because vehicles move at high speed. Therefore, the connectivity of the VANET is weak, and there arises problems such that vehicles can not communicate satisfactorily. On the other hand, road-to-vehicle communication is expected that weak connectivity of vehicle-to-vehicle communication can be improved by base station participating in communication. The Electronic Toll Collection system (ETC) [1] and the optical detector [2] that only the vehicles nearest to the RoadSide Unit (RSU) can communicate are currently being put to practical use, but in the future, it is expected that it will be possible to form networks with farther vehicles and RSUs, because a more robust ad hoc network will be formed by vehicles and RSUs relaying information.

An effective RSU placement within a limited cost is necessary since to deploy and maintain RSUs are costly. Various methods have been proposed for placement of RSUs. For the purpose of increasing communication opportunities between the vehicle and the RSUs, particularly many methods for placement the RSU based on the vehicle traffic volume have been proposed. Specifically, it is a method of placing RSUs preferentially at intersections with high traffic volume [3]. However, in order to achieve effective RSU placement, RSUs have to be deployed with the metric considering the other elements as well as the traffic volume. This is because when assuming information spreading by relaying information, in order to deploy RSU effectively, it is important to consider not only the traffic volume but also the other elements of the road network.

In this paper, we propose an RSU placement method considering road elements that affects radio signal spreading. This method consists of two actions: calculation of RSU placement priority for each intersection with considering road elements affecting radio wave spreading and operation of updating RSU placement priority. Evaluation of our proposal is performed by simulation of information dissemination by RSUs. We assumed two scenarios, which are assumed packet relay and not assumed packet relay.

In Section 2, we explain the existing methods of RSU placement. In Section 3, we describe the proposed method. In Section 4, we evaluate proposed method using simulation. We will conclude with Section 5.

II. RELATED WORK

This section explains related research on spreading information and background on RSU placement. In ITS, vehicles exchange various information, thereby improving traffic efficiency and improving safety. Various information to be exchanged here is assumed in vehicle information, such as speed and position of surrounding vehicle and itself, safety information such as where accidents occurred, entertainment information such as videos. By using this information, the vehicle can avoid the road where traffic congestion is expected and avoiding entering the area where the accident occurred.

The vehicle sends and receives these information by communication technology. Especially, ad hoc networks constituted by only vehicles are called VANETs, and various applications using VANET are assumed.

However, vehicles can not relay packets unless each vehicle are in a range of communication of each other. So, if density of vehicles is small or buildings locate where they disturb radio wave propagation, it is difficult to spread information by vehicle to vehicle. This is mentioned in papers [4], [5], [6], [7]. In a specific road structure, information spreading among vehicle to vehicle has limits because information doesn't spread wide enough by relay using Flooding. For this reason, a connectivity of vehicle to vehicle has a weak point. Then, RSU which improves the weak point of vehicle to vehicle communication are attention has been paid.

RSU is an infrastructure for telecommunication in ITS. Generally, RSU is placed on roads, and send information to vehicle and receive information from vehicle. It is assumed the situation that RSU processes information, further communicate to vehicle at local area. Furthermore, it is assumed the situation that information received by RSU are processed by server through backbone network, and RSU sends surround vehicles the packet which are processed by server. RSU is convenient but deployment cost is high. Also, it is known that cost is increased because power is frequently used for maintaining RSU once deployed. Therefore, it is not realistic to deploy infinitely many RSUs. It is necessary to effectively arrange a limited number of RSUs so that they can communicate with more vehicles.

In order to effectively utilize limited RSU, many methods of RSU placement are researched. H. Zheng et al. proposes an algorithm to set up RSU based on traffic volume to distribute advertisements at stores [8]. For each intersection, calculate the RSU placement priority high in the order of the traffic volume, and place RSUs in the place where the vehicle has many opportunities to receive the information. Similarly, J. Chi et al. also proposes an algorithm that calculates RSU placement priority at a high value based on traffic volume [3]. In addition, control is exercised not to deploy an additional RSU around the intersection where the RSU is deployed. This is an operation for preventing intersections with high RSU priority from concentrating in one place. Also, some RSU placement methods focus on deploying RSUs at highways [9], [10], [11]. Although these are references related to the placement of RSU, basically there is no constraint to deploy RSU at intersections like urban areas, so the concept of RSU placement priorities for intersections is not mentioned.

There are some analyzing methods of road network structures [12], [13], [14]. However, they don't analyze road networks from the viewpoint of radio wave diffusion. For this reason, in order to effectively deploy this RSU, we considered that it is necessary to consider the influence of static road elements in the road network on radio wave diffusion in addition to the traffic volume.

III. Proposed Method

This section proposes the RSU placement method based on road elements for information distribution. This method enables effective RSU placement suppression of redundant placement of RSU while considering the influence on radio wave diffusion by road elements. This method performs RSU placement by two operations. The first is the calculation of the RSU placement priority and the second is the update of the RSU placement priority. By calculating the priority of RSU by considering various road elements including traffic volume in the placement priority calculation of RSU, it is possible to select intersections where packets are likely to be spread as high priority intersections. Road elements are weighted based on the magnitude of influence on radio wave spread. In the update operation of the RSU placement priority, it is an operation to lower the RSU placement priority of the surrounding intersection based on the position of the existing RSU. By this operation, the redundant placement of the RSU can be suppressed, and as a result overlap of the communication range can be reduced. The p_i which express the placement priority of RSU at intersection i is expressed by the following equation:

$$p_i = w_1 \frac{1}{t_{max}} t_i + w_2 \frac{1}{s_{max}} s_i + w_3 \frac{1}{I_{max}} I_i \tag{1}$$

where t_i is a value of traffic volume at intersection i. s_i is a number of connected straight road segments at intersection i. I_i is a number of connections of the road segment at the intersection i where the connected road segment has four intersections at the other end. w_1, w_2 and w_3 are coefficients for weighting, and we set to be all $1/3$. Also, p_i is calculated between 0 and 1. The method of determining (1) is shown below.

First, in calculating the RSU placement priority, we define intersections that should be prioritized. Since information to be distributed by the RSU needs to be received by many vehicles, it is necessary to set intersections where information spreads in a wide range. Therefore, road elements that affect the spread of information to a wide range are adopted as elements for calculating the RSU placement priority. In order to spread information extensively, the information transmitted by the RSU at the intersection have to be extended over long distances and many roads. The number of road segments connected to intersections increases, information spreading tends to be effective. In addition, a straight road works favorably by spreading information because radio wave spreading is difficult to block in straight roads. Furthermore, when spreading information over a wide range with multiple hops, traffic volume becomes important because it is necessary to gather a lot of vehicles for relaying. In addition, it is conceivable that the angle of the road segments greatly influences the transmitted information to diffuse far without being blocked by buildings.

In this paper, we analyzed the correlation between road elements and the number of received cars. The road elements to be analyzed are summarized in Table. I. These road elements are classified into fields such as the number of road segments connected to the intersection, the length of the road segment, the angle formed by the road segments connected to the intersection, the angle of the connected road segment itself, the position of the intersection. For example, n_Seg denotes the number of connected road segments, d_A denotes a connection angle between road segments, and ABS 90 which is the sum of differences of each connection angle of road segments and 90°. The road map to be used this time is the map of Manhattan,

TABLE I. ROAD ELEMENTS.

classification	element
Traffic Volume	$traffic$
Number of road segment	n_Seg
	n_Inter4
	p_Inter4
Length of road segment	ave_Len_Seg
	$total\_Len\_Seg$
	d_Seg
Angle of Connection	$ABS90$
	$ABS90+$
	$ABS90+\_no\_overlap$
	$ABS180$
	$ABS180\_pair$
	$ABS180\_pair\_even$
	d_A
Angle of road segment	$n\_u-deg10$
	n_u-deg5
	n_Straight
	$p\_u-deg10$
	p_u-deg5
	p_straight
Position	$center$

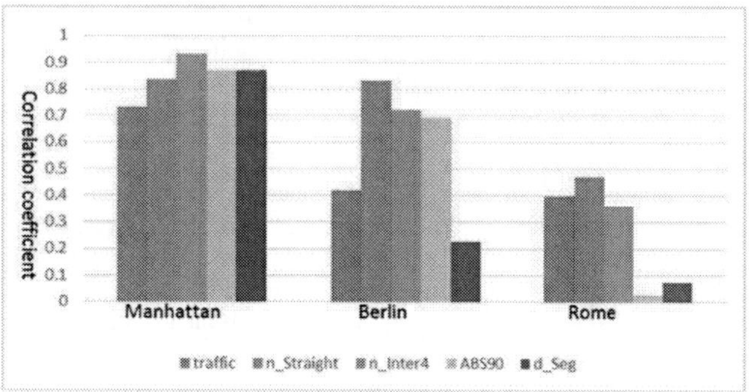

Figure 1. Correlation Coefficient between Number of Packet Received and Traffic Volume.

Berlin, Rome, $1500m$ square areas. Using these areas, we analyze road elements that affect information diffusion.

Figure. 1 represents a part of the correlation coefficient between 21 road elements and transitions of reception numbers for each intersection in each road map. In this paper, the minimum correlation coefficient to judge is 0.3. From Figure. 1, since road elements traffic, $n\_Straight$, n_Inter4 are correlated with the number of received on all road maps, these road elements are used for calculating the priority. Traffic is the traffic volume, $n\_Straight$ is the number of connected straight road segments, and n_Inter4 is the number of road segments connected with the connected road segment at the other end of the four intersections. Also, if different road elements in the same classification are duplicated for RSU priority calculation, similar road elements may duplicate and affect the RSU placement priority. Therefore, even if the correlation coefficient exceeds 0.3, only road elements with the highest correlation coefficient in the same classification shall be used for RSU placement priority calculation. Also, from the Figure. 1, even if the road elements are the same, it can be seen that the magnitude of the correlation coefficient differs depending on the road map. In addition, it can be seen that the magnitude relation with different correlation coefficients also differs depending on the road map. From these facts, it is assumed that the magnitude of the weight and the magnitude of relation between the three road elements are all equal because they change according to the road map.

The purpose of each operation is to spread the RSU widely. RSUs should be placed at intersections that are advantageous for information spreading that can distribute information to many vehicles, but if one RSU is deployed, the surrounding intersections can be covered by the RSU. In other words, if RSUs are simply placed at the intersections which are advantageous for information spreading, the RSUs are concentrated in part, the RSUs are concentrated in part and the vehicles receive duplicate and identical information from the multiple RSUs. This is called redundant RSU placement, which causes redundant information transmission. In order to perform RSU placement that uniformly transmits information to the vehicle, it is necessary to suppress the redundant RSU placement while taking advantage of the RSU placement priority. As a concrete method, it is an operation to prevent concentrating placement of the RSUs to be deployed thereafter by lowering the RSU placement priority of the intersection within the communication range of the existing RSU. Updated RSU placement priority p_{i_new} is done using (2):

$$p_{i\_new} = p_i \times \frac{dis_{RSU}}{dis_{transmit}} \tag{2}$$

where dis_{RSU} is a distance between the intersection i and the RSU deployed immediately before. $dis_{transmit}$ is the distance at which the attenuation of the priority begins to start. Since $dis_{transmit}$ plays a role of a threshold, this priority attenuation Equation is applied when the distance between the intersection i and the immediately preceding RSU is smaller than $dis_{transmit}$. In other words, it applies when this Equation $dis_{transmit}$ is above dis_{RSU}. As the value of the distance $dis_{transmit}$ at which attenuation of the priority is started this time, a value of 700 experimentally obtained is used.

A value of $dis_{istransmit}$ actually deployed the RSU in the simulation and experimentally observed and adopted a value that maximizes the number of reception. In this paper, the value of the simulation is made to conform to the reference[15] in the case where there is no specific mention. The transmission power of the radio waves of the RSU and the vehicle was set to 20 dBm, and the maximum hop number was set to 3. In the set simulation environment, the information reaches about 350 m from sender with 1 hop at 20 dBm radio field intensity. Since the maximum number of hops this time is set to 3 hops, the radio waves transmitted from the RSU can reach up to 1050 m. Therefore, we changed the hop number between 1 and 3 this time, that means $dis_{transmit}$ was 350, 700, 1050 respectively.

IV. SIMULATION EVALUATION

We evaluated the performance of our proposal by simulation. To verify the effectiveness of the proposed method, we compared our proposal with the existing methods that with existing method of RSU placement methods.

A. Simulation Set up

Here, we explain the simulation environment. In the simulation, we used two methods of RSU placement to compare with proposed method.

- Traffic 1 [8]
  An RSU placement method to calculate the placement priority of RSU at intersection based on traffic volume. There is no operation to prevent the redundant arrangement of the RSU, and simply arrange the RSU in the intersection where the traffic volume is large.
- Traffic 2 [3]
  An RSU placement method to calculate the placement priority of RSU at intersection based on traffic volume. Perform operations to prevent redundant placement of RSUs. As a specific operation, set the placement priority of RSU to a fixed distance around the already deployed RSU to 0. In this example, the distance to set the allocation priority of RSU to 0 m is set to 350 m which is the transmission range of 1 hop.

Distance operation to prevent redundant arrangement of RSUs adopted in proposed method and Traffic 2 does not necessarily work in the expected direction. This is because there is a risk that RSU placement at an intersection at which information spreading to many vehicles originally could be hindered by these distance operations.

Next, we explain about simulation scenario that we used. Place N RSUs in the simulation area. The placement of the RSU is different for each comparison target. Record simulation results when N RSUs are deployed respectively. RSUs deployed N in the area transmit packets at the same time. In the area, 500 vehicles are running at a speed of 15 to 30 km/h. 500 vehicles process received packets based on scenarios 1 and 2. In scenario 1, 500 vehicles do not relay packets received from the RSU. It is possible to receive packets only for vehicles travelling in front of RSUs. The simulation time is 600 seconds, and the RSUs in the area simultaneously transmit information 600 times in total, once a second, from simulation start to simulation end. N varies between 1 and 10. In scenario 2, 500 vehicles relay the packets received from the RSU. Even without travelling in front of the RSU, the vehicle can receive packets from the RSU within the transmission range and surrounding vehicles' relay. All vehicles participate in relaying and relay up to 3 hops. The simulation time is 130 seconds, and the RSUs in the area simultaneously transmit the information once at the time of the simulation start of 120 seconds. N varies between 1 and 20.

Finally, we explain about the simulation map that we used. The simulation area uses 1500 m × 1500 m of San Francisco. It is shown in Figures. 2. In addition, circles marks indicate RSU placement by proposed method, triangle marks indicate RSU placement by Traffic 1, and crossing marks indicate RSU placement by Traffic 2. Simulator uses Scenargie [16] and map data is acquired from OSM [17]. The packet size is 128 KB, and the radio wave propagation model uses ITU-R_P.1411 [18] which reflects the influence such as radio wave shielding by buildings. Detailed parameters are shown in the Table. II.

B. Simulation Results

Here, the results of the simulation are presented. In scenario 1, there is no packet relay by the vehicles, and it is a scenario that receives information only for vehicles passing in front of the RSU. Therefore, in scenario 1, in order to improve the reception rate, it is important to deploy RSUs at intersections where traffic volume is simply high, regardless of the radio wave spreading range.

The result is shown in Figure. 3. In San Francisco, the proposed method is the highest reception rate despite the weight of the traffic volume in the RSU calculation Equation being lower than Traffic 1 and Traffic 2. Specifically, on average, our proposal was 7.3 points higher than Traffic 2 and 27 points higher than Traffic 1. This is because San Francisco is a grid road network structure and every intersection has a similar road element. Therefore, in San Francisco, the static

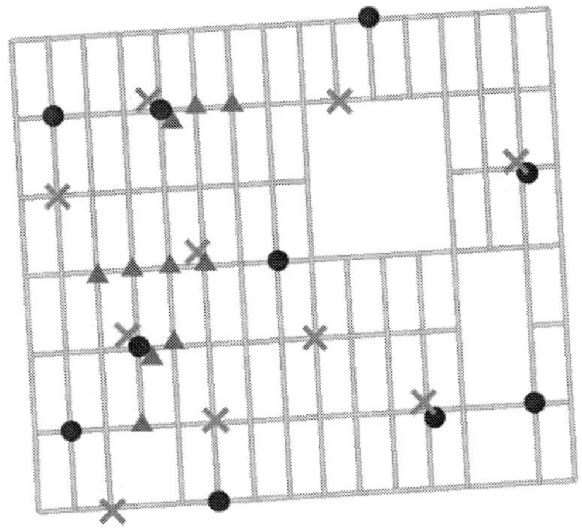

Figure 2. Simulation Area and RSU Placement in San Francisco.

TABLE II. PARAMETERS FOR EVALUATION.

Common Parameters	
Simulator	Scenargie2.0[16]
Number of vehicles	500
Velocity of vehicles	15∼30 km/h[19]
Propagation Model	ITU-R_P.1411[18]
Frequency	5.9 GHz
Bandwidth	10 MHz
Communication standard	IEEE 802.11p
Mobility Model	Random Way Point
Parameters of Scenario 1	
Simulation Time	600 s
Number of N	$1 \sim 10$
Number of Packet sent	600
Max Hop Count	1
Transmission Power	20 dBm
Parameters of Scenario 2	
Simulation Time	130
Number of N	$1 \sim 20$
Number of Packet sent	1
Max Hop Count	3
Transmission Power	20 dBm

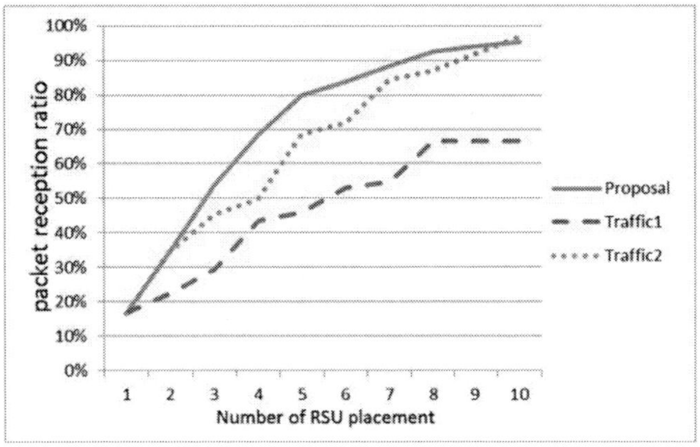

Figure 3. Reception Ratio in Scenario 1.

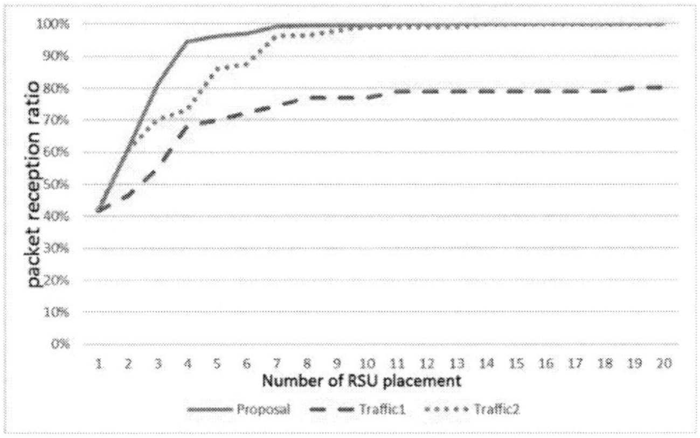

Figure 4. Reception Ratio in Scenario 2.

road elements of the road network such as n_Inter4 and $n\_Straight$ are nearly equal at every intersection, and as a result the RSU placement priority calculated in our proposal is depending on traffic volume which is a remaining element of the Equation of calculation of RSU placement. Therefore, the weights of the road elements used for the RSU placement priority calculation in the three comparison targets in San Francisco are approximately equal, and the difference in the reception ratio in this graph generated is the performance of the RSU placement priority update operation for preventing redundant RSU placement. In other words, in San Francisco the weight of the considered road element is equal for all comparison targets, so here, the performance of the updating operation of the RSU placement priority can be evaluated.

Traffic 1 that has no updating operation of the RSU placement priority has the lowest reception ratio among the comparison targets. The reception rate of Traffic 1 was about 55% even when the RSU was 10 pieces. This is thought to be due to the concentrated RSU placement because intersections whose traffic volume is high concentrate in one place. Traffic 2 suppresses redundant RSU placement by the operation of uniformly setting the priority of intersection within the communication range of RSU to 0. However, the opportunity to receive information depends on the distance from the RSU. If it ignores the distance and uniformly set the priority of the intersection within the communication range to 0 for the reason because it is within the communication range of RSU, RSU placement priority of an intersection with high possibility of working for good will is removed from candidates. Therefore, Traffic 2 has an average reception rate of 7.3 points lower than that of our proposal which linearly controls the RSU placement priority according to the distance.

In scenario 2, information relaying by the vehicle is performed. Therefore, it is important to consider the change of the radio wave spread range by the road network structure in order to increase the reception ratio. Since proposed method considers road elements related to the road network structure such as n_Inter4 and $n\_Straight$ in addition to the traffic volume to calculate the RSU placement priority, the reception rate becomes higher than Traffic 1 and Traffic 2 which consider only the traffic volume.

The result is shown in Figure. 4. As mentioned in Scenario 1, San Francisco has a regular road network structure, so there is no big difference in the road elements of each intersection. For this reason, the three comparison targets determine the RSU placement priority based on the traffic volume and RSU placement priority updating operation. The RSU placement priority updating operation works when RSUs is placed 2 or more. Therefore, when the number of RSUs is one, the three comparison targets have the same reception rate. When the number of RSUs placed is two or more, the reception ratio varies depending on the performance of the update operation of the RSU placement priority. Traffic 1 without updating the RSU placement priority is the lowest throughout the simulation graph in San Francisco. Traffic 1 converges with a difference of 20 points compared with Traffic2 and proposed method. This is thought to be due to redundant RSU placement occurring by concentrating the RSU at intersections whose traffic volume are high. Traffic 2 is an RSU placement priority updating operation of uniformly setting the priority within the communication range to 0, so the RSU placement priority updating operation does not perform corresponding to the distance between RSUs. As a result, the RSU placement priority updating operation of Traffic 1 is lower than that of proposed method, and the reception rate of traffic 1 is lower than that of the proposed method. When the number of deployed RSUs is 4, our proposal already has a reception rate of 95%, which shows that the highest reception rate is obtained throughout the graph about proposed method.

V. Conclusion

In this paper, we proposed an RSU placement method considering road elements for information distribution. This method consists of calculating operation of RSU placement priority considering road elements that affected radio wave spreading and updating RSU placement priority in response to the distance between the intersection and RSU deployed. In order to determine an operation of calculation RSU placement priority considering the road elements that influence the radio wave spreading, we analyzed the correlation with the number of 21 road elements received for the three cities of different road network structure. As the result, the number of straight road segments, the number of neighbor intersections which have 4 connected road segments and traffic volume were related to communication performance. In the simulation evaluation, we compared our proposal with existing method of RSU placement methods. In the simulation results, our proposal was the highest of the comparison targets of equivalent to the highest about the reception ratio in both scenarios with and without relay. From this result, our proposal is effective as an RSU placement method.

In the future work, we will research the value of w in (1). We use 1/3 in the equation, but we have to use accurate value to evaluate. We think w is variable in using map. So, we will research the weight of road elements in each maps.

ACKNOWLEDGMENT

This work was supported by JSPS KAKENHI Grant Number 16H02811.

REFERENCES

[1] Ministry of Land, Infrastructure, Transport and Tourism, "Trends in usage of etc," http://www.mlit.go.jp/road/yuryo/riyou.pdf, 2014.

[2] Ministry of Land, Infrastructure, Transport and Tourism, "ETC2.0(radio beacon 5.8 GHz)," http://www.vics.or.jp/know/service/etc2.html, 2015.

[3] Jeonghee Chi, Yeongwon Jo, Hyunsun Park and S. Park, "Intersection-priority based optimal rsu allocation for vanet," in proceedings of 2013 Fifth International Conference on Ubiquitous and Future Networks (ICUFN), July 2013, pp. 350–355.

[4] Ryo Yanagida, Keiji Obara, Koki Ogawa and Hiroshi Shigeno, "An analysis of road maps based on voronoi diagram for vehicular broadcast," in proceedings of 2015 10th International Conference on Broadband and Wireless Computing, Communication and Applications (BWCCA), Nov 2015, pp. 150–156.

[5] Ryo Yanagida, Keiji Obara, Koki Ogawa and Hiroshi Shigeno, "An analysis of roadmap using voronoi area for vehicular broadcast," in proceedings of IPSJ SIG technical reports ITS, May 2015, pp. 1–8.

[6] M. Fogue, P. Garrido, F. J. Martinez, J. C. Cano, C. T. Calafate and P. Manzoni, "Pawds: A roadmap profile-driven adaptive system for alert dissemination in vanets," in proceedings of 2011 IEEE 10th International Symposium on Network Computing and Applications, Aug 2011, pp. 1–8.

[7] M. Fogue, P. Garrido, F. J. Martinez, J. C. Cano, C. T. Calafate and P. Manzoni, "An adaptive system based on roadmap profiling to enhance warning message dissemination in vanets," IEEE/ACM Transactions on Networking, vol. 21, no. 3, June 2013, pp. 883–895.

[8] H. Zheng and J. Wu, "Optimizing roadside advertisement dissemination in vehicular cyber-physical systems," in proceedings of 2015 IEEE 35th International Conference on Distributed Computing Systems, June 2015, pp. 41–50.

[9] X. Liya, H. Chuanhe, L. Peng, Z. Junyu, "A randomized algorithm for roadside units placement in vehicular ad hoc network," in proceedings of 2013 IEEE 9th International Conference on Mobile Ad-hoc and Sensor Networks, Dec 2013, pp. 193–197.

[10] A. O'Driscoll and D. Pesch, "Hybrid geo-routing in urban vehicular networks," in proceedings of 2013 IEEE Vehicular Networking Conference, Dec 2013, pp. 63–70.

[11] T. J. Wu, W. Liao and C. J. Chang, "A cost-effective strategy for road-side unit placement in vehicular networks," IEEE Transactions on Communications, vol. 60, no. 8, August 2012, pp. 2295–2303.

[12] Tetsuo Ikeda, Kazumi Saito, Nobuaki Mutoh and Takayasu Fushimi, "Road network analysis by complex network analysis methods," Presentation papers of GIS Association of Japan(CD-ROM), vol. 20, 2011, pp. D–6–2.

[13] Yuta Matsumura, Tetsuo Morita and Shinya Fujita, "The study on the evaluation method of the city planning road network via graph theory," Infrastructure Planning Review, vol. 28, 2011, pp. I813–I821.

[14] Md Kamrul Hasan, Javed Iqbal Khan, Raquib Ahmed, Mollah Md Awlad Hossain and Md. Nur-Us-Shams, "Road structure analysis using gps information," in proceedings of International Conference on Electrical Information and Communication Technology (EICT) 2013, Feb 2014, pp. 1–6.

[15] Japan Automobile Research Institute, "ITScommunication simulation evaluation scenario(ver. 1.2)," http://www.jari.or.jp/Portals/0/resource/pdf/H23_simyu/%EF%BC%88-Ver1.2%EF%BC%8920131010.pdf, 2012.

[16] Yasunori Owada, Taka Maeno, Shigeru Kaneda, Ryosuke Hisanaga and Mineo Takai, "Realistic ITS Simulation with Scenargie Simulator," in proceedings of IPSJ DICOMO, 2008, pp. 233–234.

[17] H. Mordechai and W. Patrick, "OpenStreetMap: User-Generated Street Maps," IEEE Pervasive Computing, vol. 7, no. 4, 2008, pp. 12–18.

[18] ITU-R P.1411-7, "Propagation data and prediction methods for the planning of short-range outdoor radiocommunication systems and radio local area networks in the frequency range 300 MHz to 100 GHz," International Telecommunication Union, Recommendation, 2013.

[19] Ministry of Land, Infrastructure, Transport and Tourism, "Travel speed of the three major cities," http://www.mlit.go.jp/road/ir/ir-data/data/107.pdf, 2010.

Remote Proxy V2V Messaging Using IPv6 and GeoNetworking

Masahiro Kitazawa*, Manabu Tsukada*, Kai Morino†, Hideya Ochiai* and Hiroshi Esaki*
*Graduate School of Information Science and Technology
The University of Tokyo
Email: {ktzw, tsukada, jo2lxq}@hongo.wide.ad.jp, hiroshi@wide.ad.jp
†Institute of Industrial Science
The University of Tokyo
Email: morino@sat.t.u-tokyo.ac.jp

***Abstract*—Vehicle-to-vehicle (V2V) messaging plays an important role in cooperative intelligent transportation systems (CITS), which are advanced applications addressing the problems of road transport management. In this regard, the Cooperative Awareness Message (CAM) protocol is standardized in EU for V2V messaging. However, V2V messages work well only when all vehicles have transmitters and work well in short range and no obstacle. To solve the former problem, the existance of the non-ITS road users, in the early deployment phase, we previously proposed a system called Proxy CAM, wherein roadside sensors detect target vehicles and transmit V2V messages on behalf of the vehicles. However, the V2V transmission range is still limited to the wireless range of IEEE802.11p. Therefore, in this study, we propose a system that delivers CAMs over the Internet using UDP/IPv6 and LTE in addition to the standard specification (i.e., the Basic Transportation Protocol/GeoNetworking and IEEE802.11p). Moreover, we implement the cellular network component of the system and evaluate its performance in terms of the packet delivery ratio and packet delay for various distances and packet frequencies. In our evaluation, we define the average Proxy CAM update delay and calculate this parameter for both IEEE802.11p and LTE. We find that using LTE over long distances is more efficient than using IEEE802.11p.**

Keywords*–*CITS; CAM; LTE; Cellular Network; V2V.

I. Introduction

The problem of vehicular traffic, particularly traffic jams and accidents, has become one of the most important issues in the world today. In this context, intelligent transport systems (ITSs) have been designed to solve this problem via acquiring and sharing traffic information from vehicle sensors or other devices. Examples of ITS include Vehicle Information and Communication Systems (VICS) [1], which automatically receive information about traffic jams and road construction from roadside beacons via FM multiplex broadcasting and electronic toll collection systems (ETCSs) [2] installed at the entrances and exits of highways by communicating with vehicles via radio waves. In particular, ITSs that communicate traffic information with other ITSs to improve vehicle safety, durability, efficiency, and comfort are called cooperative ITSs (CITSs). A typical CITS consists of an application layer, facilities layer, network&transport layer, access layer, management layer, and security layer [3].

A CITS communicates by means of Cooperative Awareness Messages (CAMs) [4] and Decentralized Environmental Notification Message (DENM) [5] protocols. A CAM contains information on the positions and movements of road users (vehicles, bicycles, pedestrians) along with other dynamic information. According to [4], all ITS-installed devices are encouraged to create and broadcast messages with the use of CAM, and the desirable frequency range for CAM broadcasting is 1 to 10 Hz. Further, single-hop broadcasting is also desirable. Upon receiving CAM information, a CITS processes the CAM to update the Local Dynamic Map (LDM) [6]. The Cooperative Awareness basic service (CA basic service) in the facilities layer, the Basic Transport Protocol (BTP) [7]/GeoNetworking(GN) [8] in the network&transport layer, and IEEE802.11p [9] in the access layer are used to send and receive CAMs. The CA basic service also manages to encode and decode CAMs and provide CAM information to the LDM and application layer, while BTP provides end-to-end connectionless communication. This protocol is aimed at transmitting multiple messages from different processes in the facilities layer at the same time along one packet path. However, it is to be noted that BTP does not guarantee the order, integrity, and reliability of packets. Meanwhile, the GN protocol provides the packet path while IEEE802.11p specifies the wireless communication system developed for communication between vehicles. This makes communication between fast-moving vehicles possible in the frequency band of 5.85∼5.925 GHz. The bandwidth is narrow compared to other IEEE802.11 series, and therefore, the communication speed is less, but resistance to multipath propagation is strong.

A CITS cannot function without communicating with other vehicles; however, there are certain problems with the current communication protocols. First, the communication protocols do not cover non-CITS road users such as vehicles with no CITS and pedestrians. In this regard, it is noteworthy that there are no CITS products commercially available today in Japan. Thus, the commercialization of CITS vehicles requires the system functionality to detect neighboring non-CITS vehicles and pedestrians. While computer vision can be utilized to solve this problem, there is a possibility that computer vision cannot detect vehicles in the case that they are in blind spots. There are always blind spots at intersections, and vehicles may not be able to avoid accidents with other vehicles or pedestrians emerging from such blind spots. Second, IEEE802.11p uses the frequency band, 5.85∼5.925 GHz. This range of frequencies is so high that these waves hardly undergo diffraction; thus, they cannot travel around obstacles. Finally, the signal strength of such wireless radio systems also decreases with increasing distance.

In the above context, this study proposes the usage of cellular networks (UDP/IP) and LTE in addition to the original Proxy CAM [10] system's protocol stack, BTP/GN + IEEE802.11p, to solve the abovementioned problems. In our study, we design a Remote Proxy CAM system that uses computer vision sensors installed along the roadside to generate Proxy CAMs from the acquired images and broadcasts them using BTP/GN + IEEE802.11p (as does the original Proxy CAM system). In addition, the Remote Proxy CAM system uses a server-client model to send request-based Proxy CAMs to the client.

In the study, we also implement a prototype of a part of this system using a cell phone, and we measure the latency and delivery ratio of the packets. We find that packets using UDP/IP + LTE exhibit short delay times and high delivery ratios and function efficiently over long distances.

The contributions of this work are:

- Analysis of problems of Proxy CAM
- Proposal of a communication protocol stack in addition to the original one
- Prototype implementation of a part of the proposed system
- Experimental evaluation of the prototype

The rest of the paper is organized as follows. Section II highlights related works, and Section III analyzes the issues of Proxy CAM and summarizes the requirements of the solution. Section IV presents the design of our system along with its implementation. Section V demonstrates and evaluates the implementation, and finally, Section VI concludes our paper and presents our future studies in this direction.

II. Related Works

In this section, we introduce and discuss two main related concepts, Proxy CAM and cloud-based pedestrian road-safety. The Proxy CAM approach is used to generate CAMs by proxy while cloud-based pedestrian road-safety involves communication between vehicles and pedestrians using a cellular network.

A. *Proxy CAM*

In general, when CITS cars broadcast their CAM information in the vicinity of obstacles such as buildings, there is a possibility that the CAMs cannot reach cars the other side of the buildings, which may lead to accidents. Further, the presence of non-CITS cars (which cannot broadcast their CAM information) cannot be detected by CITS cars without computer vision. To solve this problem, we propose the Proxy CAM [10] system that works with any vehicle-sensing technology. This system consists of a roadside sensor, sensor fusion database, Proxy CAM generator, and Proxy CAM transmitter. In our scheme, first, roadside sensors detect vehicles and acquire their position, speed, acceleration, and other optional information. Next, the sensors tag each vehicle with an ID that is sent to the sensor fusion database along with the other information. Any type of road traffic sensor can be utilized with the system if it can detect vehicles and acquire the relevant information. The sensor fusion database receives the vehicle data and stores it in the sensor fusion local dynamic map (SFLDM). This database also has the functionality to identify a single vehicle from the data from multiple sensors and integrate the data. Next, the Proxy CAM generator uses the SFLDM and composes Proxy CAMs. Some of the CAM fields are labeled as 'unknown'. The Station ID field of the proxy CAM contains 24 bits set to '1' and 8 random bits. If the sensor or SFLDM identifies the vehicle information as that of a previously detected vehicle, the Station ID field of its Proxy CAM is set to the same value as the one created previously. Finally, the Proxy CAM transmitters broadcast the generated Proxy CAMs using the ETSI standard protocols, BTP/GN in the networking&transport layer, and IEEE802.11p in the access layer. The CAM transmitters should be installed at locations ensuring a clear line-of-sight to the vehicles.

B. *Cloud-Based Pedestrian Road-Safety*

Pedestrians can be detected only with the use of computer vision by CITS cars because pedestrians cannot create and broadcast CAMs. However, the accuracy of computer vision depends on the weather and daylight conditions (day or night). Thus, to accurately detect the presence of pedestrians, we need another method. In this regard, in a previous study [11], the authors used the cell phones of pedestrians to communicate with CITS cars. However, CITS cars can only communicate using IEEE802.11p while cell phones do not use IEEE802.11p. Thus, the authors of the study installed cell phones in CITS cars, and the phones were connected to the CITS. However, in cellular networks, broadcasting is not allowed. Thus, the authors used a cloud server to communicate with pedestrian and the CITS cell phones. In the scheme, both sets of cell phones (vehicular and pedestrian) always send their positions to the server. The pedestrian's cell phone communicates at low frequencies because the communication consumes electrical power and the capacity of the cell phone battery is limited. The server always verifies the positions of both the pedestrian and CITS phones. If the two parties approach each other, the server sends a request to the pedestrian's cell phone to switch the communication frequency to 'high'. After receiving the request, the pedestrian's cell phone switches to the designated high frequency. If the server estimates the possibility of a collision between the pedestrian and car, it sends an alert message to both of them.

C. *Other Research*

Our approach is also inspired by several other relevant studies, which we briefly present here. In [12], the authors used power lines to communicate between ITS users and the transport management system. In [13], Vehicular Ad-Hoc Networks (VANETs) were utilized for communication, and the authors designed two VANET sampling protocols named SAME and TOME to collect vehicular traffic information and detect incidents in real-time. Further, [14] used parked cars to forward CAMs from cars on the opposite sides of a building.

III. Problem Statement

In this section, we discuss the problems of Proxy CAM in detail. Subsequently, we analyze the design requirements for the solution.

A. *Problem*

Figure 1 illustrates the operation of Proxy CAM and the problems of the system. The Proxy CAM device 1) detects the target vehicle and 2) delivers the proxy CAM to the receiver over IEEE802.11p. However, the system suffers from two significant problems, as described below.

1) Limited wireless range: As indicated by label a) in Figure 1, we observe that a message cannot be delivered beyond the wireless range. According to [15], if a car drives at 60 km/h, IEEE802.11p's packet delivery ratio (PDR) is nearly 100% for a received signal strength indicator (RSSI) value of -85 (about 800 m). However, the authors performed their experiments on level land in the absence of buildings and other electromagnetic interference; on the other hand, urban environments contain buildings and other forms of electromagnetic waves, which can lead to packet delivery failure beyond certain distances. Further, [10] have reported that the packet reach distance is about 60∼70 m if there are buildings in the environment. Moreover, a CITS vehicle at 60 km/h cannot stop before an intersection. Thus, it is necessary to widen the coverage of Proxy CAM transmission.

2) Interference and Obstacles: Theoretically, Proxy CAMs can be received at roads connected to intersections. However, if a car is in a situation where another obstacle blocks the line of sight to the CAM transmitter, as indicated by label b) of Figure 1, it cannot receive Proxy CAMs. Further, if there is a vehicle such as a truck between a car and Proxy CAM device, the car cannot receive Proxy CAMs. Therefore, it is necessary to ensure that the Proxy CAMs surmount such obstacles.

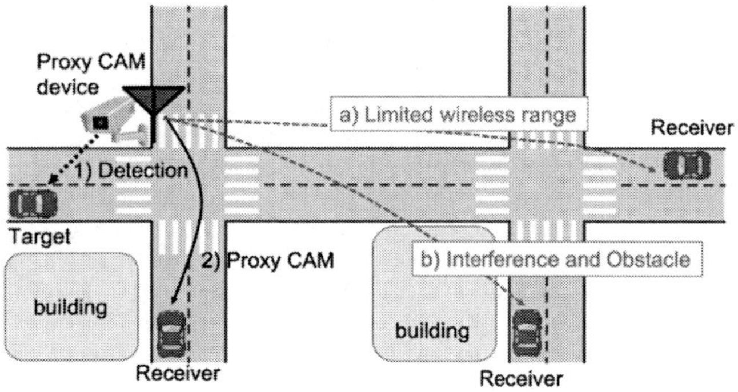

Figure 1. Overview and problems of Proxy Cooperative Awareness Message (CAM).

B. Requirements

To solve the abovementioned problems, we designed a system that uses a new protocol stack in addition to the original protocol stack, BTP/GN + IEEE802.11p. The following section lists the requirements of the new system.

1) Message Transmission Coverage: In urban environments, CAMs should be able to cover a range of distances. Our system can send CAMs to everyone in the coverage area regardless of the presence of buildings or large vehicles.

2) Availability: A CAM contains information regarding neighboring cars. Thus, the unavailability of a CAM for even 5 s can lead to an accident. Consequently, a system should allow CAMs to be sent any given time. Importantly, the communication should be stable (i.e., the PDR should be nearly 100%).

3) Real-Time Information: The position of a vehicle is dynamic, and therefore, CITS cars should always obtain updated information reflecting the real-time situation. That is, the time between the creation of a CAM and its reception should be as short as possible. Moreover, delays in sensing and message transmission must be minimized.

4) Using existing protocol and wireless communication system: For interoperability among countries, CITSs are developed based on a given architecture, protocols, and technologies. From the perspective of practical application, the system should not use new resources; it should consist of existing protocols and use wireless communication.

IV. Remote Proxy CAM

To satisfy the abovementioned requirements, we designed the Remote Proxy CAM.

A. System Design

In our system, we use LTE and the cellular network to satisfy the communication requirements. Since broadcasting is not allowed in cellular networks, we use unicast communication to access the dynamic vehicular information. We use IPv6 for the Remote Proxy CAM because it fulfills the CITS requirements through its extended address space, embedded security, enhanced mobility support, and ease of configuration. It also uses UDP because the delivered message comprises real-time data. By using UDP/IPv6, the packet can be transmitted over the LTE and a cellular network. Figure 2 shows the protocol stack of the proposed method. The vehicle with our proposed system receives Proxy CAMs via BTP/GN + IEEE802.11p as well as UDP/IPv6 + LTE.

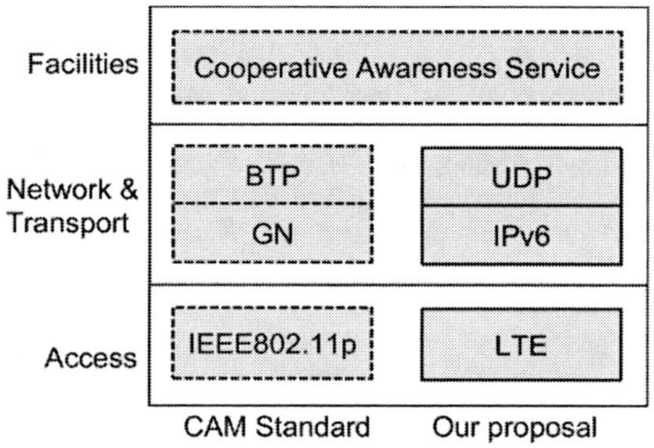

Figure 2. Protocol stack of Proxy Cooperative Awareness Message (CAM).

Figure 3 shows the overview of the proposed method. First, a roadside computer vision sensor detects vehicles around the intersection and creates the corresponding CAMs. Next, the sensor broadcasts the CAMs. At the same time, the vehicle sends a Remote Proxy CAM request to the Proxy CAM device along the vehicle route as per the demand of the ITS application. When the Proxy CAM device receives a request message from a vehicle via the cellular network, it sends a Proxy CAM reply to the vehicle via the cellular network. If the vehicle receives Proxy CAMs via both IEEE802.11p and LTE, it updates the LDM entry with the newest information.

Here, we remark that the IP address discovery of the Proxy CAM device is out of the scope of the paper. Possible solutions for the IP address discovery include embedding the IP address in the digital map, downloading the static list of IP addresses, or resolving the IPv6 address from the geographical information by means of a DNS-like system.

1) Vehicle: A vehicle sends a request message for CAMs to the nearest Proxy CAM device along the vehicle route via the cellular network using UDP/IPv6 and LTE. If the vehicle does not receive a CAM response from the Proxy CAM devices after a given interval, it sends a request again. The vehicle always checks for CAMs using IEEE802.11p or LTE, and if it receives CAMs, it updates its LDM with the newest ones.

2) Proxy CAM device: The created CAMs are broadcasted using standard protocols based on ISO and ETSI. Our broadcast uses IEEE802.11p in the access layer and BTP and GN in network&transport layer. At the same time, the created CAMs are transmitted to vehicles that have sent requests for CAMs in the unicast via the cellular network. This communication uses LTE in the access layer and UDP and IPv6 in the network&transport layer. Further, this transmission continues for a specified interval, after which it is terminated. The transmission is resumed if the Proxy CAM device again receives a request.

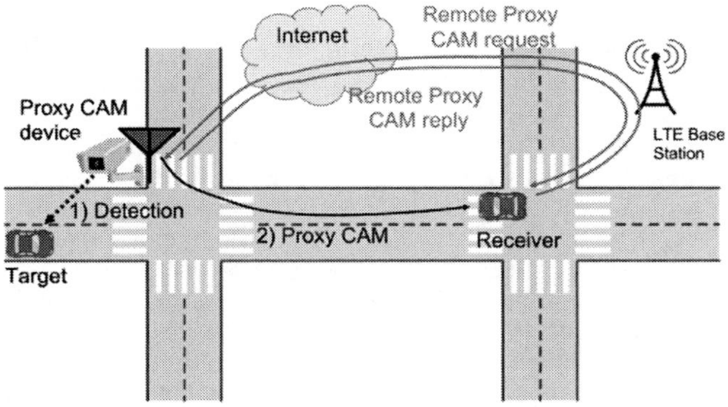

Figure 3. Overview of Remote Proxy Cooperative Awareness Message (CAM).

B. System Implementation

We implemented the cellular network part of our designed system as shown in Figure 4. We used the cellular device, REI (FTJ161B-REI) manufactured by FREETEL, in our study. We tethered the cellular device and the receiver through a USB cable. Our program was written in C language.

1) Transmitter: We used the LGN-20-00 transmitter manufactured by Commsignia Ltd. as the packet transmitter. The transmitter communicated with the router via a 50-m-long Ethernet cable in the access layer (Figure 4(a)). The router and receiver were connected via a 30-m-long Ethernet cable, and the transmitter was configured to send packets through the cable in order to measure the reference value of the delay (Figure 4(b)). These two sets of communications used UDP/IPv6 in the network&transport layer. We used multicast for pathway (a) and unicast for pathway (b) in Figure 4. The transmitter used a single program to send the same packet along the two paths. In order to realize the Pub/Sub model based on which the Proxy CAM device sends CAMs via the cellular network after receiving a request, the transmitter begins to function after receiving a request packet via pathway (a). The IP address of the destination in pathway (a) is the sender IP address in the request packet. As regards pathway (b), we set the global IP address of the receiver interface in advance.

2) Router: In our study, we used the PR-400NE manufactured by NTT as the router. The router and receiver were connected by means of a 35-m-long Ethernet cable. The firewall did not filter any kinds of IPv6 packets.

3) Reciever: We used the Tier PC Note GTX970M as the packet receiver. The processors in the device include the Intel Core i7-4720HQ CPU @ 2.60 GHz×8, with a memory capacity of 31.3 GB and the OS being Ubuntu 15.04. We tethered the receiver and cellular device through a USB cable. We used two programs: one to receive the packets via pathway (a), and the other to receive the packets via pathway (b). To realize the Pub/Sub model, we ensured that the receiver sent a request packet first.

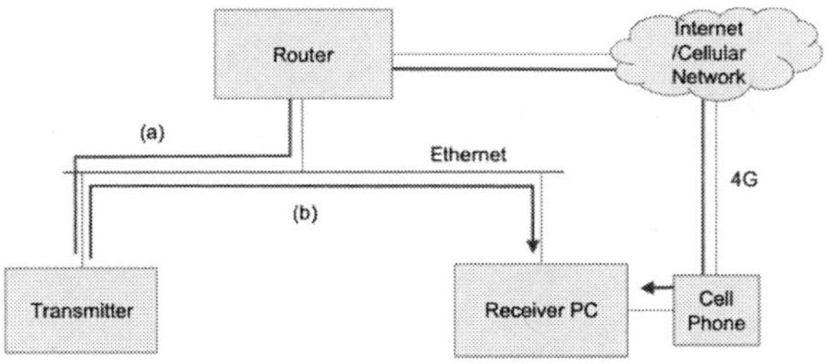

Figure 4. Network configuration.

V. EVALUATION

We evaluated the implementation of our proposed method in terms of the PDR and delay. We performed our experiments in a residential area. In the experiment, first, the transmitter is set to create sockets for pathways (a) and (b) in Figure 4 and to wait for the CAM request via pathway (a). When the transmitter receives the request, it sends out a given number of packets at a predetermined frequency. The size of each packet is 300 bytes, and there is a number in a certain packet that identifies it as the first packet. Next, the receiver creates sockets for pathways (a) and (b) and files to record the result. When the receiver receives a packet, it obtains the reception time using the `gettimeofday` function from `sys/time.h` and records the time to the file. After our experiment, we compared the reception time of the same number packets of the two files and measured the delay obtained using pathway (a).

We performed our experiments at frequencies of 1, 5, 10, 50, 100, 500, 1000 Hz and distances of 10, 30, and 50 m. To ensure that the PDR was more than three significant figures and the experiment time more than 60 s, we set the total number of packets as 100 for 1 Hz, 300 for 5 Hz, 600 for 10 Hz, 3000 for 50 Hz, 6000 for 100 Hz, 30000 for 500 Hz, and 60000 for 1000 Hz.

A. Packet Delivery Ratio (PDR)

The following table lists the PDR results of our experiments. From the table, we note that the PDR hardly depends on the frequency and distance. The maximum packet loss rate is 0.08%. In theory, it can also be said that the distance between the transmitter and the receiver does not matter because communication occurs between a cellular device and a base station. This experiment demonstrates that communication between the transmitter and receiver using the cellular network, UDP/IPv6 + LTE, is possible nearly 100% of the time regardless of the frequency and distance.

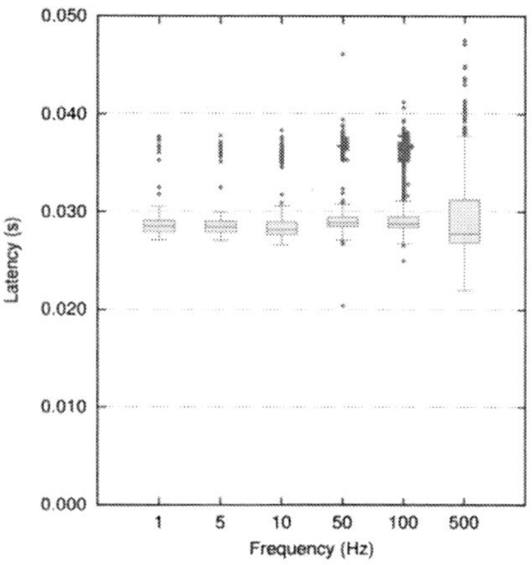

Figure 5. Packet delay (10 m).

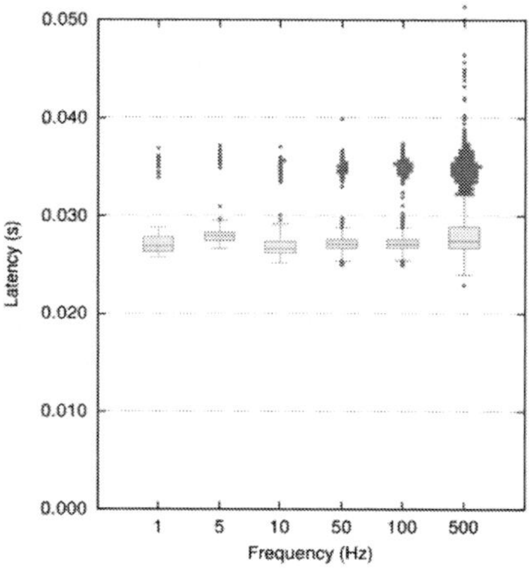

Figure 6. Packet delay (30 m).

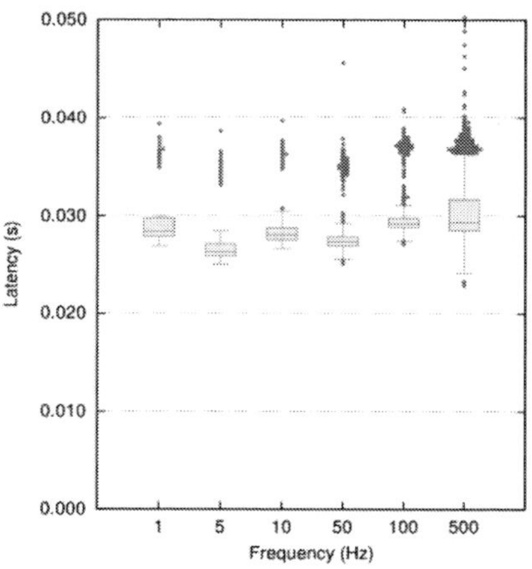

Figure 7. Packet delay (50 m).

TABLE I. NUMBER OF RECEIVED PACKETS FOR VARIOUS DISTANCE WITH USE OF UDP/IPv6 + LTE.

	10 m	30 m	50 m
1 Hz (100 pkts)	100 pkts	100 pkts	100 pkts
5 Hz (300 pkts)	300 pkts	300 pkts	300 pkts
10 Hz (600 pkts)	600 pkts	600 pkts	600 pkts
50 Hz (3000 pkts)	3000 pkts	3000 pkts	3000 pkts
100 Hz (6000 pkts)	6000 pkts	6000 pkts	6000 pkts
500 Hz (30000 pkts)	29987 pkts	29980 pkts	29976 pkts

B. *Packet Delay*

Figures 5,6, and 7 depict the packet delay for different frequencies and distances. When we performed the experiment at 1000 Hz, the transmitter could not generate packets every 0.001 s. Therefore, we have excluded the result for 1000 Hz. The overall results indicate that the delay does not depend on the distance; it is constant at about 30 ms. As mentioned in the PDR section, the distance between the transmitter and receiver does not matter in theory.

Next, we attempt to further understand the implications of our results. First, to evaluate our results, we defined the misregistration delay as the time from the transmission of a CAM from a Proxy CAM device to the reception of the **next** CAM at the vehicle. This parameter indicates the difference in the position of the vehicle in terms of the CAM and the position of the vehicle in reality. We defined this parameter as the sum of the CAM reception interval and the packet delay. The reception interval can be calculated as the product of the CAM transmission interval and the reciprocal of the PDR. We defined the PDR as p_{PCAM}, frequency of CAM transmission as f_{PCAM}, and packet delay as $t_{PCAM-trans}$. Consequently, the misregistration was calculated using Equation (1).

$$\frac{1}{p_{PCAM}} \times \frac{1}{f_{PCAM}} + t_{PCAM-trans} \tag{1}$$

We calculated the misregistration delay from our results. We assumed the frequency of the CAM transmission as 10 Hz for the maximum desirable CAM broadcast frequency of 10 Hz according to [4]. We set the PDR to 1 because the results indicate that the PDR did not depend on the frequency and distance; further, the PDR was nearly 1. We also set the packet delay to 0.03 seconds. With these values, the misregistration delay was estimated as 0.13 seconds. We did not perform experiments using BTP/GN + IEEE802.11p, since it is not legal for application outside Japan. Therefore, we applied the result of the IEEE802.11b/g protocol as per the field test of [10]. In this case, the PDR was 90%, 80%, and 75% for distances of 10 m, 30 m, and 50 m, respectively. For the packet delay, the distance between the transmitter and receiver was a maximum of 50 m, and the electromagnetic signals were considered to travel instantaneously, i.e., the time of signal travel was 0 s. For these abovementioned values, the misregistration delay was estimated as 0.11 s, 0.126 s, and 0.133 s for distances of 10 m, 30 m, and 50 m, respectively. Upon comparing the misregistration delay of UDP/IPv6 + LTE with that of BTP/GN + IEEE802.11b/g, we concluded that UDP/IPv6 + LTE and BTP/GN + IEEE802.11b/g perform equivalently in terms of the performance speed.

C. *Analysis on intersection scenario*

Here, we discuss how our proposed method improves pedestrian and vehicular safety at intersections with respect to vehicle distance and approach speed towards an intersection. Let us consider the situation that a vehicle is approaching an intersection with an obstacle close by. To avoid collision with the obstacle, the vehicle must stop before it enters the intersection. In this study, we analyzed the distance required for the vehicle to stop before entering the intersection for the initial speed of the vehicle. We denoted the distance between the initial position of the vehicle and the intersection as d, its deceleration as m, which is assumed constant during the scenario, initial vehicle speed as v_0, and time interval from the instant of consideration as t. The safe distance to avoid collision can be calculated using Equation (2) as follows:

$$d = \int_0^{v_0/m} v(t)dt = \frac{1}{2} \cdot \frac{v_0^2}{m}, \tag{2}$$

where $v(t)$ denotes the speed of the vehicle at time t and $v_0 = v(0)$ the initial speed of the vehicle. Figure 8 depicts this safe distance d as the black solid curve obtained with m set to 1.0 m/s^2.

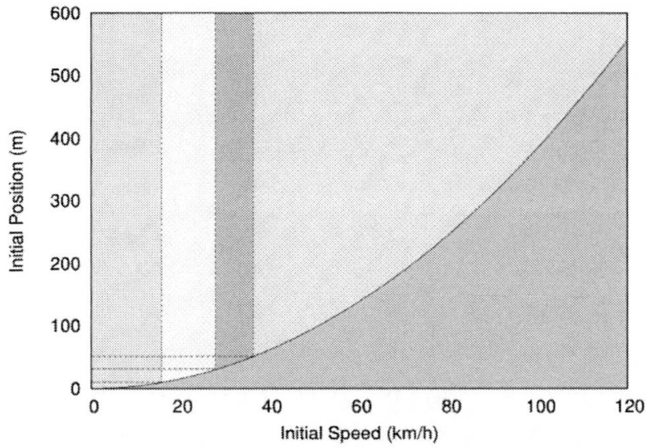

Figure 8. Safe stopping distances for various initial distances to intersection and initial vehicle speeds.

In this figure, the region under the black curve (colored in red) indicates the unsafe distance-and-speed range, i.e., the vehicle is moving too fast, and it cannot stop before entering the intersection. The other four regions lying above the curve correspond to speeds and distances over which the vehicle can stop before entering the intersection if it receives a CAM or the driver sights the object/obstacle beforehand. There are four possible collision-prevention solutions corresponding to receiving a CAM or sighting the obstacle. Here, we summarize the effective distance for each solution. The first solution corresponds to sighting of the obstacle by the human eye or computer vision, which works well if the obstacle can be noticed. In the situation that the obstacle is in a blind spot, effective preventive action can only be taken when the vehicle is very close to the obstacle. Therefore, we set the maximum effective (communication) distance for this case as 10 m. The second solution corresponds to the use of the original CAM system. According to Figure 5(b) of [10], the maximum communication distance for such a system is 30 m. The third solution considers our Proxy CAM system, wherein the maximum distance depends on the situation. In the absence of buildings and other obstacles, CAMs can travel over 800 m([15]). However, the effective braking distance is constrained by the presence of buildings and other large vehicles such as trucks or buses. Thus, we set the maximum distance to 50 m for the Proxy CAM case. The final solution considers the efficacy of our proposed method. This method uses both a cellular network and LTE. Thus, in theory, the system can work anywhere on the road (except in long tunnels). Therefore, its maximum communication distance can be considered as ∞m.

Each of the abovementioned effective distances determines the safety margins of the intersection with obstacles nearby. Thus, the safety regions can be classified into the four regions shown in Figure 8. Human eyes and computer vision are the poorest solutions; their use can prevent an accident if the vehicle's position and speed lie in the orange region. The original CAM system can prevent an accident if the vehicle's position and speed lie in the orange and yellow regions. Meanwhile, the Proxy CAM system can prevent collision if the vehicle's position and speed lie in the orange, yellow, and blue regions. The green region is still unsafe for all three solutions. However, the proposed method can still prevent accidents corresponding to the green region because our system can communicate regardless of the distance. Therefore, our proposed method is highly effective over existing technologies.

VI. Conclusion and Future Work

We proposed a system called the Remote Proxy CAM that uses a cellular network with UDP/IPv6 + LTE in addition to the original Proxy CAM system protocol stack, BTP/GN + IEEE802.11p, in order to widen the coverage of the CAM transmission and improve failure tolerance. To evaluate this system, we implemented the cellular-network component of the system and performed various experiments. In the experiments, the transmitter sent packets to the receiver with various frequencies over different transmitter-receiver distances. Our results indicated that using the cellular network with UDP/IPv6 + LTE afforded a high PDR (nearly 100%) and low average delay of about 30 ms, which indicate that the proposed method is stable and operates in real-time. This method allows CITS vehicles to stably and consistently receive CAMs of faraway vehicles in real-time.

We are planning to focus on three aspects of the system in our future works. Currently, we have assumed that the vehicle knows the position and IP address of the Proxy CAM device; however, our system needs to have a discovery mechanism for discovering the proxy CAM device. Further, we need to examine possible solutions for device discovery, such as IP-address-embedded digital map, download-based solution, or a DNS-like system, as mentioned in section IV-A. Second, we require a system that compares the created time of CAMs across both IEEE802.11p and LTE and then updates LDM with the newest times. Finally, we performed our experiments with one receiver while actual situations would require more receivers. Thus, the proposed method should be tested with multiple receivers and the signal delay must be investigated.

Acknowledgement

The authors would like to thank Ye TAO and Xin LI for their help with the field tests and various comments on the paper. This work was supported by JSPS KAKENHI Grant Numbers JP17H04678 and JP26730045.

References

[1] Vehicle Information and Communication System, URL: http://www.vics.or.jp/know/about/pdf/vics_pamphlet_j.pdf, Retrieved: Jun, 2017.

[2] Outline of ETC2.0, URL: http://www.go-etc.jp/etc2/etc2/index.html, Retrieved: Jun, 2017.

[3] Draft ETSI EN 302 636-3 v1.1.2(2014-03),Intelligent Transport Systems (ITS); Vehicular Communications; GeoNetworking; Part 3: Network Architecture, 2014.

[4] ETSI EN 302 637-2 v1.3.2(2014-11), Intelligent Transport Systems(ITS); Vehicular Communications; Basic Set of Applications; Part 2: Specification of Cooperative Awareness Basic Service, 2014.

[5] ETSI EN 302 637-3 v1.2.2(2014-11), Intelligent Transport Systems(ITS); Vehicular Communications; Basic Set of Applications; Part 3: Specifications of Decentralized Environmental Notification Basic Service, 2014.

[6] ETSI EN 302 895 v1.1.1(2014-09), Intelligent Transport Systems(ITS); Vehicular Communications; Basic Set of Applications; Local Dynamic Map(LDM), 2014.

[7] Intelligent Transport Systems (ITS); Vehicular Communications; GeoNetworking; Part 5: Transport Protocols; Sub-part 1: Basic Transport Protocol, 2011.

[8] Intelligent Transport Systems (ITS); Vehicular communications; GeoNetworking; Part 4: Geographical addressing and forwarding for point-to-point and point-to-multipoint communications; Sub-part 1: Media-Independent Functionality, 2011.

[9] IEEE Standard for Information technology-Telecommunications and information exchange between sy tems-Local and metropolitan area networks-Specific requirements Part 11: Wireless LAN Medium Access Control(MAC) and Physical Layer(PHY) Specifications Amendment 6: Wireless Access in Vehicular Environments, 2010.

[10] T. Kitazato, M. Tsukada, H. Ochiai, and H. Esaki, "Proxy cooperative awareness message: an infrastructure-assisted v2v messaging," in 2016 Ninth International Conference on Mobile Computing and Ubiquitous Networking (ICMU), Oct 2016, pp. 1–6.

[11] M. Bagheri, M. Siekkinen, and J. K. Nurminen, Cloud-Based Pedestrian Road-Safety with Situation-Adaptive Energy-Efficient Communication, 2016, URL: http://ieeexplore.ieee.org/document/7518724/, Retrieved: Jun, 2017.

[12] C. A. Sakurai, C. L. Marte, L. R. Yoshioka, and C. F. Fontana, "Integrating intelligent transportation systems devices using power line communication," in Recent Researches in Telecommunications, Informatics, Electronics and Signal Processing, 2014, pp. 36–42.

[13] A. Baiocchi, F. Cuomo, M. D. Felice, and G. Fusco, "Vehicular ad-hoc networks sampling protocols for traffic monitoring and incident detection in intelligent transportation systems," in Transportation Research Part C: Emerging Technologies Volume:56, Mar 2015, pp. 36–42.

[14] D. Eckhoff, C. Sommer, R. German, and F. Dressler, "Cooperative awareness at low vehicle densities: How parked cars can help see through buildings," in 2011 IEEE Global Telecommunications Conference - GLOBECOM 2011. IEEE, Dec 2011, pp. 1–6.

[15] O. Shagdar, M. Tsukada, M. Kakiuchi, T. Toukabri, and T. Ernst, "Experimentation towards ipv6 over ieee 802.11p with its station architecture," in International Workshop on IPv6-based Vehicular Networks, Jun 2012, pp. 1–6, URL: https://hal.inria.fr/hal-00702923, Retrieved: Jun, 2017.

Towards Vehicle-Assisted Adaptive Wireless GeoMesh Network for Smarter Cities

Yin Chen*, Takuro Yonezawa* and Jin Nakazawa†

*Graduate School of Media and Governance
†Faculty of Environment and Information Studies
Keio University, Endo 5322, Fujisawa, Kanagawa, Japan
Email: {yin,takuro,jin}@ht.sfc.keio.ac.jp

***Abstract*—Recent development in low-power wireless networks enables us to deploy IoT (Internet of Things) devices in an entire city with low maintenance and management cost. However, since such IoT devices cannot have fine-grained location sensors, such as GPS, types of applications with the IoT devices must be limited. This paper proposes a method of fine-grained location estimation for IoT devices in low-power wireless networks, specifically Wi-SUN (Wireless Smart Utility Network). We leverage public vehicles, which go around over an entire city almost everyday with mounting a GPS sensor and Wi-SUN link. The vehicular devices are used to collect GPS data for learning the relationships between the radio strength measured at Wi-SUN stations and the locations of IoT devices. With the learned data, we expect to estimate the fine-grained location of a Wi-SUN device.**

Keywords–Wi-SUN; Localization; Vehicular Network.

I. Introduction

The increasing concentration of the world's population in urban areas places the cities in a spotlight. Indeed, on 2% of the earth's surface, cities actually use 75% of the world resources [1]. These aspects inevitably make the cities important actors for the world's sustainable development strategy. Especially in developed countries including Japan, problems on people (e.g., continuousely increasing aging population and weakened links of communities) and problems on city infrastructure (e.g., worn roads, bridges and buildings) in urban areas are considered as one of the biggest social problems. Moreover, cities must have tenacious and dependable functionalities against big disasters, such as earthquakes or typhoons. To solve these problems, ICT (Information and Communication Technologies) have to be more integrated and leveraged to support urban life for citizens, municipalities and businesses. Especially, IoT technology and networks to support IoT are promising ways to enhancing both awareness of cities' contexts and responsiveness of activities.

Recent development in wireless ad-hoc network technologies offers several practical protocols, such as WiFi, Bluetooth, Zigbee, Wi-SUN [2], LoRa [3] and SIGFOX. Compared to networks as fixed infrastructure, ad-hoc networks have advantages of flexibility and adaptability. Especially, in this research, we focus on low-power communication protocols, such as Wi-SUN and LoRa, because low-power consumption is one of the key requirements for easy maintenance and management of IoT deployed in a city. To get the benefit of low-power consumption, IoT devices should not have a high power-consuming location sensors, such as GPS (Global Positioning System). However, fine-grained location information is indispensable to support various smart cities applications.

In this paper, we propose a fine-grained location estimation method in Wi-SUN mesh networks. To achieve the goal, we leverage public vehicles to collect fingerprint of radio strength measured at Wi-SUN stations and ground-truth GPS coordinates for location estimation. Public vehicles, such as garbage trucks, move around over an entire city almost everyday. Therefore, we can obtain timely relationships between radio strength and GPS locations, and adapt the relationship for Wi-SUN devices to estimate locations accurately. The contributions of the paper are as follows:

- Addressing the problem for fine-grained location estimation in wireless mesh network with low-power consumption protocols.
- Proposing a method to improve the location estimation accuracy by leveraging public vehicles.

The rest of the paper are organized as follows. Section II introduces Wi-SUN mesh networks and the location estimation problem. We describe our proposed approach for location estimation in Section III. Section IV makes a brief review of related work. Finally, we conclude the paper in Section V.

II. Location Estimation Problem in Wireless Mesh Networks

This section firstly introduces Wi-SUN, the adopted wireless communication protocol. Then, we discuss the location estimation problem in Wi-SUN mesh networks.

A. *Wi-SUN*

Wi-SUN, short for Wireless Smart Ubiquitous Network, is a wireless specification using IEEE802.15.4g in PHY layer and IEEE802.15.4e in MAC layer. Wi-SUN is designed to meet the requirements for M2M/IoT communication, such as low cost, low power consumption, autonomous operation. Compared to other LPWA (Low Power and Wide Area) protocols, such as LoRaWAN or SIGFOX, Wi-SUN supports up to 300 Kbps bandwidth, and forms not only star topology but also mesh topology called Wi-SUN FAN (Field Area Network). These features allow for multiple and redundant connection paths with high network performance compared to other LPWA protocols, so that it provides a reliable IoT network service for smart cities.

In our project, we will deploy a wireless mesh network with Wi-SUN FAN in Fujisawa city, Japan. In Fujisawa, like in all the other cities in Japan, emergency municipal radio communication stations are installed about every 200 meters for public announcement of emergency information, such as evacuation. We integrate those stations and Wi-SUN FAN networks to provide a communication network covering the entire city. Figure 1 maps the locations of all the radio communication stations in Fujisawa city. The communication range of Wi-SUN is around 500 meters, therefore, deploying Wi-SUN base stations to the stations forms a practical wireless mesh network over the entire city.

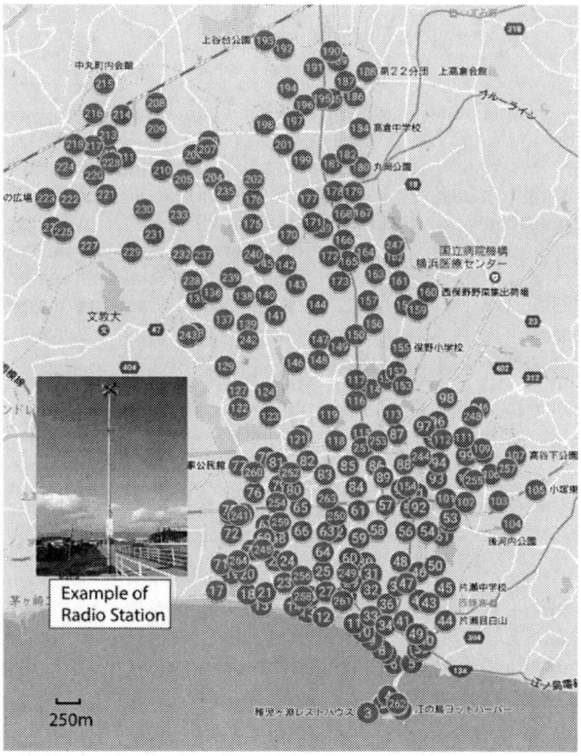

Figure 1: Locations of emergency municipal radio communication stations in Fujisawa city, Japan

B. *Location estimation in Wi-SUN mesh network*

The location is an important information for many application scenarios in Wi-SUN and other LPWA protocols. One of the typical use cases is to watch over elderly people of dementia; when an elderly is abnormally wandering away from home, a sensor device with Wi-SUN beacon attached to him, will detect the abnormality and let care-workers and families know his locations. Similar applications are also useful for young children or pets.

Usually, the location of a device in Wi-SUN network is estimated through identifying which Wi-SUN station is closest to the device. However, this only estimates the device's location as a radius of about 200 meters. This granularity is too wide when targeting people to be watched over for protection. To estimate accurate locations, triangulation can be adapted by measuring Wi-SUN device's RSSI (Received Signal Strength Indication) at multiple Wi-SUN stations, the locations of which are known. However, since signal strength varies greatly in different environmental conditions (e.g., weather, season and so on), detailed relationships between RSSI and the locations should be updated timely. This limitation makes the estimation of accruate locations very difficult, because there is no suitable model to updating the relationship in an entire city. We propose an approach to solve the problem in the next section.

III. Improve Estimation with Public Vehicles

In this section, we first describe the architecture of the proposed system and then introduce the idea of vehicle-assisted location estimation and the automotive sensing platform to be adopted in our research, respectively.

A. *Overview of the proposed system*

As shown in Figure 2, we will implement in this research a system that consists of self-developed Wi-SUN edge devices, Wi-SUN base stations and vehicular devices with GPS receiver and Wi-SUN module. The Wi-SUN edge devices will be attached to users, e.g., young children, the elderly and pets. The base stations will be deployed to emergency municipal radio communication stations of Fujisawa city. The devices broadcast signals periodically or when requested by base stations. If a base station is in the vicinity of a edge device, it will be able to measure the RSSI of the edge device. Via the collaboration of multiple base stations, it is envisioned that locations of the edges devices can be estimated via triangulation. The vehicular devices will be installed into garbage trucks of Fujisawa city and collect GPS data of their locations to assist the location estimation of base stations.

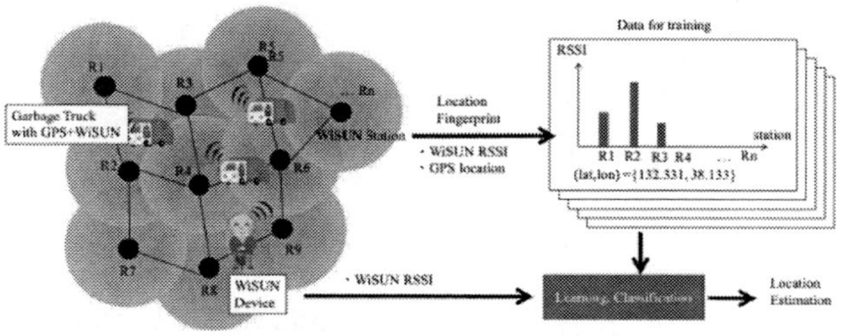

Figure 2: System overview

B. *Vehicular-assisted location estimation*

In outdoor environment, the propagation of wireless signals may be affected directly or indirectly by weather, seasons, buildings and so on, so that the wireless signal strength (RSSI) would be variable over time. In this research, we propose to use public vehicles, like buses, taxi cars or garbage trucks, equipped with GPS and Wi-SUN module, to enhance the accuracy of location estimation at the base stations. In particular, these vehicles will broadcast their locations, i.e., GPS coordinations, as they move around in the city. The base stations in the vicinity of these vehicles will receive these GPS data and also measure the corresponding RSSI of these signals transmitted from the trucks. The GPS data and RSSI data will be further analyzed to find the relationship between locations and wireless signal strength, which in turn can be used to improve the location estimation of our system. It is notable that since public vehicles typically roam around over their cities almost everyday, so timely update can be expected.

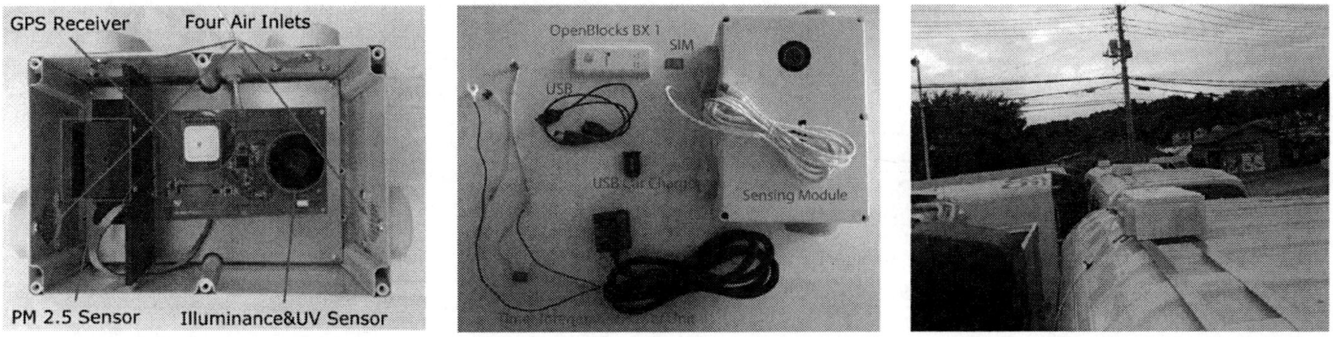

(a) (b) (c)

Figure 3: Sensor node of Cruisers.

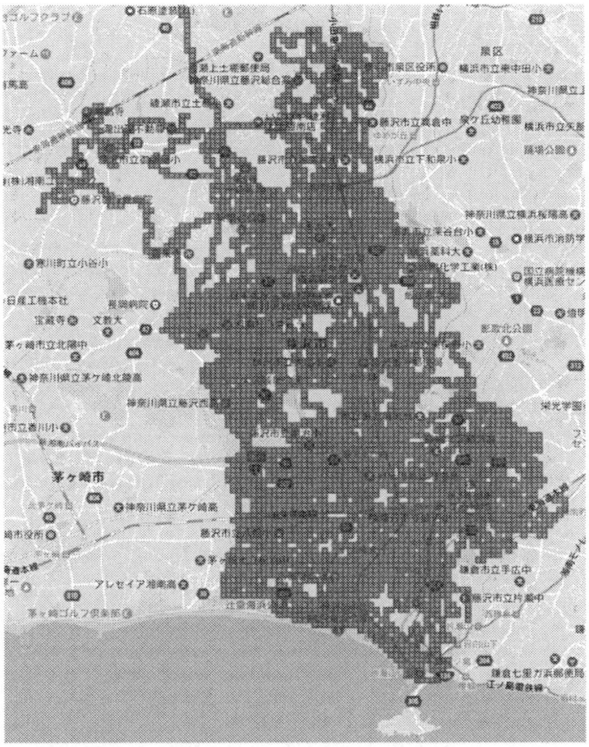

Figure 4: The coverage of Cruisers. The size of a cell is set to $100m \times 100m$. A cell is marked with red color if a truck of Cruisers visits the grid at least once per three days.

C. *Cruisers: an automotive sensing platform*

In our past research, an automotive sensing platform called Cruisers has been deployed to garbage collection trucks in Fujisawa. The Cruisers platform will be used in this research to serve as the host of vehicular devices. As shown in Figure 3, a sensing module containing a GPS receiver and environmental sensors are developed and installed on the roof of garbage trucks. The sensor data are read by the small computer with embedded 3G connectivity, called OpenBloks BX1, and sent to backend servers over a carrier network.

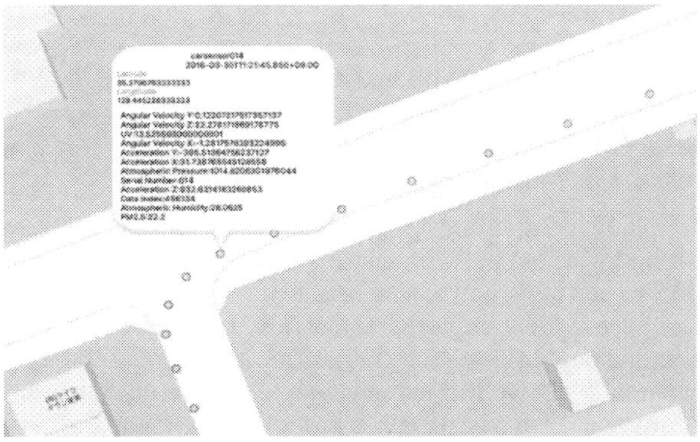

Figure 5: Fine-grained sensing of Cruisers

The city operates 140 garbage trucks for waste collecting and 66 of them have been equipped with the platform by June, 2017. The garbage trucks provide a good coverage on Japanese cities, since the garbage collecting system in Japan works in a daily and door-to-door fashion. In Japan, each house/mansion maintains at least one garbage container located in a nearby roadside. Following the schedule made by local government, residents will throw their family waste to the container in the early morning of each *collecting day* (i.e., the days, typically weekdays, that garbage collecting trucks come to collect waste). Meanwhile, the garbage collecting trucks operated by local governments are organized to visit these containers successively to collect house waste.

Based on the GPS data accumulated from Cruisers, it has been shown that a wide coverage can be achieved [4]. Figure 4 shows the Cruisers' coverage performance. In Figure 4, the urban area of the city are divided into a grid of $100m \times 100m$. A cell is marked with red color if a truck of Cruisers visits the grid at least once per three days. Since we haven't fully deployed the platform to all the garbage trucks, we mainly cover the southern part of the city. The Cruisers-attached trucks visit all the residents' houses in the southern part, implying it covers the area where residents' activities are held. Additionally, with the platform, a fine-grained data collection can be achieved as shown in Figure 5. The sensory data are sent to the backend servers in 100Hz. Supposing a truck is driving at 40km/h, the 100Hz sensing will give us a set of sensor data for each 11.1cm.

IV. Related work

The recent progress in ICT like IoT, 5G and LPWA has inspired a number of wireless mesh network applications in smart cities. In [5], the authors proposed a wireless mesh sensor network based on WiFi and discussed its applications in smart grid, agriculture and environment protection. While WiFi provides a high bandwidth, its high energy consumption hinders its application in power-limited scenarios. A routing protocol was proposed in [6] to use Bluetooth devices to construct wireless mesh networks. Simulation was also conducted in [6] to show how the proposed protocol outperforms conventional routing schemes like blood routing in terms of energy consumption. A data collecting scheme using frame aggregation of Wi-SUN was proposed in [7] to improve the data aggregation efficiency of a WSN of smart meters. It is notable that compared with mature wireless networking technologies, like WiFi and cellular communication, LPWA technologies like Wi-SUN, LoRa and SIGFOX achieve long range communications as well as low power consumption, making them promising options for smart cities.

Node positioning is a fundamental problem in power-limited wireless networks where a GPS module is not practical due to its high power consumption. While node positioning has been widely explored in wireless sensor networks [8] and cellular networks [9], node positioning in Wi-SUN networks remains a not well investigated area. On the other hand, WiFi-based approach has also been proposed to enhance GPS service in indoor environment [10]. It is envisioned that while the methods and experience obtained from previous research are helpful, the wireless radiation in the proposed network will be more sensitive to environment factors like weather, so that new estimation methods are required to address these issues. Finally, how to effectively utilize vehicular nodes to improve the estimation accuracy is also an interesting problem.

Automotive sensing is a novel sensing technology where sensors are installed into vehicles to utilize vehicles' mobility to conduct urban sensing. Besides our work [4][11], a couple of applications have also been proposed in the literature. In [12], the authors equipped nuclear radiation sensors to agricultural vehicles to detect nuclear pollution on farmland at Fukushima, where a nuclear leakage accident occurred after the Great East Japan earthquake. The authors of [13] used smartphones stood on the dashboard of a car to detect good view of flowering cherries along roadside. An on-demand air quality monitoring system using public vehicles was proposed in [14], where the authors studies how to adapt the monitoring behavior of sensors in order to accumulate sufficient information of air with a relatively few number of vehicles.

V. Conclusion

The paper propose a method to estimating fine-grained locations in Wi-SUN mesh networks. To solve unstable RSSI in outdoor environment, we propose to leverage sensorized public vehicles for collecting relationships between RSSI and GPS locations. By providing accurate locations, various applications will be inspired. As the first step, we have developed an automotive sensing platform called Cruisers. We will implement and delpoy all the components of the system, and evaluate its location estimation accuracy and feasibility in Fujisawa city. Through our research, vehicles should play an important role not only in wireless mesh network but also to support fine-grained location estimation. Thus, our study will open up a new networking model of vehicle-integrated functional city networks to support various smart city applications.

Acknowledgment

This research and development work was supported by the MIC/SCOPE #171503013.

References

[1] J. von Uexküll and H. Girardet, Shaping our future: Creating the World Future Council. Green Books, May 2005, URL: http://digital.library.unt.edu/ark:/67531/metadc13722/ [retrieved: 2017-06-24].

[2] "Wi-SUN Alliance," 2017, URL: https://www.wi-sun.org/index.php/en/ [retrieved: 2017-06-24].

[3] "LoRa Alliance," 2017, URL: https://www.lora-alliance.org [retrieved: 2017-06-24].

[4] Y. Chen, T. Yonezawa, J. Nakazawa, and H. Tokuda, "Evaluating the Spatio-temporal Coverage of Automotive Sensing for Smart Cities," in 2017 Tenth International Conference on Mobile Computing and Ubiquitous Network (ICMU) (ICMU2017), 2017, Manuscript submitted for publication.

[5] L. Li, H. Xiaoguang, C. Ke, and H. Ketai, "The applications of wifi-based wireless sensor network in internet of things and smart grid," in 2011 6th IEEE Conference on Industrial Electronics and Applications (ICIEA). IEEE, 2011, pp. 789–793.

[6] H.-S. Kim, J. Lee, and J. W. Jang, "BLEmesh: A Wireless Mesh Network Protocol for Bluetooth Low Energy Devices," in 2015 3rd International Conference on Future Internet of Things and Cloud. IEEE, 2015, pp. 558–563.

[7] F. Kojima and H. Harada, "Superframe division multi-hop data collection with aggregation on Wi-SUN profile for ECHONET Lite," in 2014 IEEE Wireless Communications and Networking Conference Workshops (WCNCW), 2014, pp. 116–121.

[8] X. Ji and H. Zha, "Sensor positioning in wireless ad-hoc sensor networks using multidimensional scaling," in INFOCOM 2004. Twenty-third AnnualJoint Conference of the IEEE Computer and Communications Societies, vol. 4. IEEE, 2004, pp. 2652–2661.

[9] J. Schloemann, H. S. Dhillon, and R. M. Buehrer, "Toward a tractable analysis of localization fundamentals in cellular networks," IEEE Transactions on Wireless Communications, vol. 15, no. 3, 2016, pp. 1768–1782.

[10] S. Bell, W. R. Jung, and V. Krishnakumar, "WiFi-based enhanced positioning systems: accuracy through mapping, calibration, and classification," in Proceedings of the 2nd ACM SIGSPATIAL International Workshop on Indoor Spatial Awareness, 2010, pp. 3–9.

[11] Y. Chen, J. Nakazawa, T. Yonezawa, T. Kawasaki, and H. Tokuda, "An Empirical Study on Coverage-Ensured Automotive Sensing using Door-to-door Garbage Collecting Trucks," in Proceedings of the 2nd International Workshop on Smart Cities: People, Technology and Data, 2016, pp. 6:1–6:6.

[12] Y. Matsumoto and M. Satoh, "Tablet-type GPS tracking radiation detection system and viewer software," in IEEE SENSORS 2014 Proceedings, 2014, pp. 229–232.

[13] S. e. Morishita, "SakuraSensor: Quasi-realtime Cherry-lined Roads Detection Through Participatory Video Sensing by Cars," in Proceedings of the 2015 ACM International Joint Conference on Pervasive and Ubiquitous Computing, 2015, pp. 695–705.

[14] Y. Shirai, Y. Kishino, F. Naya, and Y. Yanagisawa, "Toward On-Demand Urban Air Quality Monitoring Using Public Vehicles," in Proceedings of the 2nd International Workshop on Smart Cities: People, Technology and Data, 2016, pp. 1:1–1:6.

Data Routing Challenges in UAV-assisted Vehicular Ad hoc Networks

Hakim Ghazzai, Hamid Menouar, and Abdullah Kadri
Qatar Mobility Innovations Center (QMIC)
Qatar University
Doha, Qatar
e-mails: {hakimg, hamidm, abdullahk}@qmic.com

Abstract— **Reliable and efficient transportation system in smart cities relies on effective integration of connected and autonomous vehicles. Although there has been a lot of improvements in this technology, additional support from the emerging unmanned aerial vehicles (UAVs) are needed to streamline the future transportation systems' capabilities. This paper investigates the problem of data routing in UAV-assisted vehicular ad hoc networks (VANETs) composed of multiple flying nodes and intelligent cars aiming at exchanging messages through the dynamic network topology. The introduction of UAVs can offer multiple advantages to the routing process thanks to their free mobility and enhanced channel quality but, at the same time, they are subject to many limitations including their limited batteries and positioning issues. This paper provides a comprehensive survey about the advantages and challenges related to the use of UAVs in data routing in VANETs. Then, it discusses some existing routing protocols for connected vehicles supported by UAVs. Finally, the paper introduces the idea of combining routing protocols with UAV locations and/or path adjustment solutions to enable efficient data routing for delay-tolerant applications.**

Keywords- Connected vehicles; data routing; intelligent transportation systems; unmanned aerial vehicle.

I. INTRODUCTION

Intelligent transportation system (ITS) constitutes one of the most important key components of future smart cities [1]. Reliable and efficient ITS is necessary to ensure safe and smarter transport networks. This requires the integration of multiple technologies, such as advanced wireless communication, computational, and sensing techniques to exchange and collect accurate and real-time information about the traffic and state of the transport network. Connected vehicles play an essential role in rendering the driving experience safer and more efficient. Using embedded communication capabilities, the vehicles can autonomously inform their neighbors about their status and motion parameters to mitigate any danger in the roads and avoid traffic jams [2]. The collected information can also be forward via cellular networks or deployed road side units (RSUs) to the traffic operator in order to monitor and control the transport networks.

Research in connected vehicles has witnessed significant advances over the last decade [3]-[5]. The most tackled issue in such scenarios is the mitigation of the instability of the wireless links connecting these mobile vehicles. Several routing algorithms have been proposed in the literature optimize the exchange and collection of data in these vehicular ad hoc networks (VANETs). However, the performances of connected vehicles remain limited due to the short ranges of the communication links, their random mobility, and fast dynamic topology change [6]. Therefore, there is still a need to support the intelligent vehicles with other road and transportation components. To this end, micro-unmanned aerial vehicles (micro-UAVs), aka drones, can be efficient candidates with great potentials to support the ground vehicles to overcome their limits and improve the quality of service of diverse ITS applications. Hence, UAV-assisted VANETs are henceforth the leading solution for data exchange and collection in ITS.

In this paper, we investigate the data routing process in UAV-assisted VANETs. In Section II, we will highlight the advantages and challenges related to the employment of UAVs in ITS. Afterwards, in Section III, a brief survey summarizing the developed data routing techniques with an emphasis on the UAV mobility impact is presented. Next, in Section IV, we investigate a particular scenario where we propose our joint routing path selection and UAV positioning solution. Finally, we conclude the paper in Section V.

II. ADVANTAGES AND CHALLENGES OF UAVS IN DATA ROUTING

Supporting connected vehicles by UAVs offer potential gains to enhance the performance of data exchange. UAVs can play an important role in routing the data between the different ground nodes due to the following advantages:

• **Channel Quality:** Thanks to their placement flexibility especially at high altitudes, UAVs can provide reliable line-of-sight (LoS) communication links with the ground vehicles [7]. Indeed, the higher the altitude is, the more the probability to establish LoS links occur. In this way, the system throughput of this air-to-ground (A2G) or ground-to-air (G2A) link is enhanced allowing the transmission of higher amount of data during the routing process. Moreover, transferring data from UAV to UAV is much more efficient than using traditional V2V communications through ground-to-ground (G2G) channels. Indeed, air-to-air (A2A) channels are subject to lower path loss and shadowing effects, which offer better channel quality. Hence, exchanging data between two ground nodes through multiple flying UAVs is much more efficient than using traditional methods. Mathematically, the free-space path losses in dB of A2A and G2G links corresponding to a LoS and non-LoS (NLoS) channels can be written as follows:

$$PL^{A2A} = PL^{LoS} = 10\gamma \log_{10}\left(\frac{4\pi f_c d}{C}\right) + L^{LoS}, \quad (1)$$

$$PL^{G2G} = PL^{NLoS} = 10\gamma \log_{10}\left(\frac{4\pi f_c d}{C}\right) + L^{NLoS}, \quad (2)$$

where γ is the path loss exponent, f_c is the carrier frequency, d is the Euclidean distance separating the two nodes, C is the speed of light, and L^{LoS} and L^{NLoS} are two additional attenuation terms for the LoS and NLoS environments, respectively, such that $L^{NLoS} \gg L^{LoS}$. It should be noted that other path loss models can be adopted.

Regarding the A2G links, their path losses can be written as a linear function of the LoS and NLoS path losses weighted by the probability of having a path loss connection denoted by p_{LoS}. The A2G path loss is written as follows:

$$PL^{A2G} = P_{LoS} PL^{LoS} + (1 - p_{LoS}) PL^{NLoS}. \quad (3)$$

An expression of the path loss probability p_{LoS} has been derived in [8] as follows:

$$p_{LoS} = \frac{1}{1+\alpha \exp(-\beta(\theta(h,d)-\alpha))}, \quad (4)$$

where α and β are constants which depend on the environment and θ denotes the elevation angle and is given by $\theta = \frac{180}{pi} \sin^{-1}\left(\frac{h}{d}\right)$ where h is the UAV's altitude. From this expression, we can deduce that UAV placed at high altitude will increase the probability of having a LoS link and hence, improving the channel condition. At the same time, it may increase its distance with ground node. A tradeoff between the altitude and the distance separating the communicating has to be achieved to enhance the channel quality.

• **Free Mobility:** Thanks to their three dimension (3D) mobility, UAVs can offer additional degrees of freedom to enhance the data routing. Indeed, they can cover larger areas with the ability to transmit collected data in real-time or store it on-board for future use. Moreover, unlike vehicles that have to move according to road directions or fixed road side units (RSU), UAVs can be placed at any location in order to establish direct connectivity with other nodes. This flexibility allows these flying nodes to connect other out-of-range nodes and act as relays for their communications. Therefore, placing UAVs in optimized locations and/or efficiently planning their paths would contribute in enhancing the data routing process in UAV-supported VANETs. This constitutes one of the major challenges in data routing through the flying nodes. In addition, to the traditional path selection task that has to be performed in the VANET dynamic topology, placing and/or moving the UAVs to ensure better support to the routing path represents another objective that has to be jointly optimized. In such scenarios, balancing between path selection using traditional routing protocols and UAV positioning is a non-trivial task mainly for moving UAVs.

The free mobility can also be subject to certain limitations. First, UAVs are not necessarily employed to route data. In most of the cases, they are used for other tasks, such as traffic monitoring or collecting images or videos. Hence, they can participate in the data routing procedure, i.e., the secondary task, if the objective of the primary task is not affected. Therefore, the support offered by the UAVs might be limited in space and time. For instance, UAVs cannot freely move out of certain regions defined by the operator or related to the primary task. Moreover, relaying UAVs may not be always available due to their limited primary task engagement or allocated energy budget. Finally, the UAV mobility has a negative effect on the channel quality. Indeed, high Doppler spread can be observed when both the UAV and car are mobile. This may limit the performance of the routing process.

• **Battery-Limited Nodes:** Unlike vehicle nodes, UAVs are battery powered and require frequent to-and-fro trips to reload their batteries, which may affect their contributions in the data routing process [9]. Hence, the energy consumption of UAVs has to be taken into account during the data routing optimization procedure to ensure seamless communication between the nodes. In addition to the energy of the communication interface, additional and relatively more important energy is consumed to ensure the hovering and forward flight of the UAVs. Hence, the UAVs that are selected to participate in the routing procedure need to have sufficient energy to complete the data transfer. This may impact the transmit power level of the UAVs, affect their communication range, and limit their mobility.

III. Data Routing Techniques and Use-cases

In UAV-assisted VANET, UAVs and ground vehicles can contribute to the data routing procedure where messages can be exchanged through air, ground, or both. The path selection is dependent on the objective of the process, such as the reduction of the total transmission time or the conservation of the link stability. Some of the applications in ITS allow some delay in the data transfer. Hence, the data routing procedure can be adapted to the type of applications which can be classified to delay-intolerant applications and delay-tolerant applications. In the sequel, we briefly discuss the routing protocols that can be applied in UAV-assisted VANET and the challenges corresponding to the tolerance of applications.

A. *Routing Protocols*

Several routing protocols involving UAVs have been discussed in literature [10]-[12]. Most of them are applied to Flying Ad hoc Networks (FANET) only. The routing protocols are classified as proactive, reactive, hybrid, and geographic routing protocols. The first protocol category assumes either fixed routing tables loaded to the UAVs before operation or periodically refreshed tables where the latest updates are considered to transfer the data. These protocols assume low topology variation even if the UAVs are in motion. The use of fixed routing tables requires a fixed topology. This can be applied for a group of nodes (UAVs and/or ground vehicles) having coordinated movement which is not the regular case in UAV-assisted VANET. For low frequencies of routing tables update, the protocols' performances can significantly degrade due to the very dynamic topology of the network and the distributed control of the nodes and their mobility. Optimized link state routing protocol (OLSR) is one of the prominent algorithms that are used in VANETs and FANETs. However, several extensions of OLSR have been presented in the literature to cope with the nodes mobility and high signaling overhead problems. In [13], directional antennas and cross-layer schemes based on flight information of UAVs have been added to the functionalities of the traditional OLSR in order to enhance its routing table's updates. A fast OLSR protocol has been proposed in [14] to meet the need of highly dynamic topology. This leads to a considerable increase in the signaling overhead especially for the in-motion node. The authors of [12] have proposed a predictive OLSR protocol using Global Positioning System (GPS) information to measure the quality of wireless links between nodes. The movement of UAVs are predicted based on their past mobility and speed. The routing tables are then pre-constructed based on these predictions. The performances of all these OLSR extensions remain dependent on the routing tables' accuracy and the stability of the links connecting the nodes.

The second routing protocol category known as reactive protocols proposes to discover paths for data transfer on demand. With such protocols, periodic messages are avoided but a delay has to be considered in order to establish the routing path before the data transmission. In addition to that, these protocols are designed for wireless mesh networks where a source needs to find a routing path to transfer its data to a target destination. Ad-hoc On-demand Distance Vector (AODV) is one of these reactive protocols where the source is aware about the next-hop information only [15]. Hybrid protocols represent a combination of proactive and reactive protocols. This protocol category will lead to more efficient performances but may cause additional delays in order to discover route in addition to extra signaling overhead.

The geographic routing protocol is a position-based routing protocol assuming that each node is aware of its neighbors' geographic locations [16]. The data transfer is performed such that each node selects its closest neighbor to the destination such that the distance to destination is reduced. In some cases, this greedy behavior of the algorithm fails to find a next hop closer to the destination. In such a scenario, face routing can be applied to help in finding another route to be followed by the geographic routing protocol. Although these greedy protocols do not require routing tables, their achieved routing paths remain suboptimal and may require higher number of hops before reaching destination compared to other categories.

B. *UAV Path and/or Location Adjustment for Data Routing*

The aforementioned algorithms, which are originally designed and applied to MANETs and VANETs, are applied for delay-intolerant applications where the objective is to find a route as fast as possible to deliver data. The total routing operation time, which is consisted of the route construction time and the data transfer time, is in the order of milliseconds. In such scenarios, the routing process requires the existence of communication links between some of the nodes such that the data can be routed. However, in many cases and especially in ITS, seamless routing paths might not be found due to deficient communication links between two or multiple nodes. Therefore, shifting the locations of some nodes can enhance the link quality and allow the discovery of routing paths. In UAV-assisted VANET, dedicated UAVs can be controlled by the operator to ensure connectivity. The operator may decide to place UAVs in precise locations to act as relays for other ground nodes. Another scenario could correspond to the case when relaying is not the principle task of UAVs. In such cases, UAVs can modify their current locations or their initial paths if they are in motion to support other nodes. For on-demand applications, the path or location adjustments of UAVs can lead to additional delay in the order of seconds and possibly minutes due to the flying time corresponding to the UAVs shift. Hence, such routing process involving UAVs path or location adjustment can be applied for delay-tolerant applications.

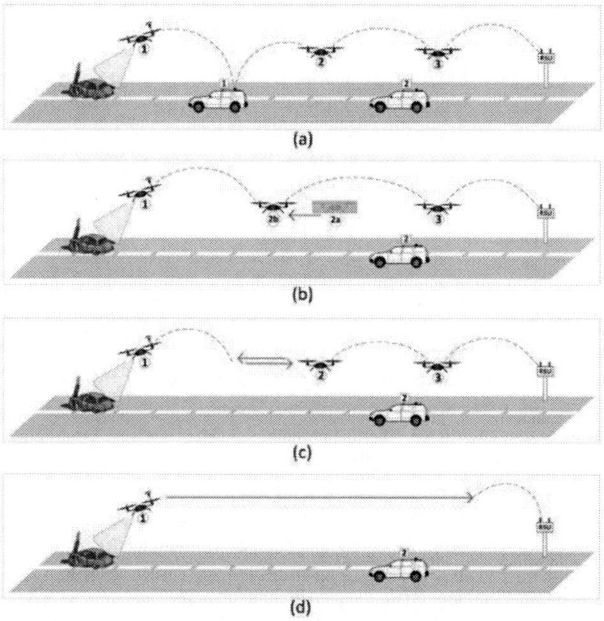

Figure 1. Examples of routing scenarios in UAV-assisted VANETs.

In Figure 1, we illustrate examples of some routing scenarios involving UAVs. In Figure 1(a), the UAV collects information about a particular event in the transport network. The data is routed through flying and ground vehicles already deployed in the area as direct communication links exist. Delay-intolerant applications require the existence of such topology to enable successful routing. Notice that the absence of ground vehicle 1 causes the failure of the routing process due to long distance separating UAV 1 and UAV 2. Hence, in order to overcome this issue and enable successful routing, one possibility based on location adjustment is shown in Figure 1(b). In this case, UAV 2 shifts its location from position (2a) to position (2b) in order to establish a direct link with UAV 1 without breaking up its link with UAV 3. This location adjustment will certainly cause certain delay in the routing due to the motion of the UAV but remain a good solution for applications tolerating a certain delay. If the distance separating UAV 1 and UAV 3 is very long such that it is not possible to find a location for UAV 2 where it can establish direct communication links with its peers, UAV 2 seeks close location to UAV 1 to collect the data and then, return next to UAV 3 to forward it as shown in Figure 1(c). Finally, another special case of routing is given in Figure 1(d). One or multiple UAVs act as data collectors where each UAV has to follow a particular path to collect and store necessary information from various locations then, go back and send the data to the sink. Many use-cases related to this type of applications exist in ITS. In the following, we cite few of them:

- **Flying Accident Report Agent:** In this case, one or a set of UAVs fly to the accident's location to get a detailed report about the accident. The collected information need to be sent to the related authorities. When reaching the accident's location, the UAV can exploit the ground vehicles to transfer the data. Other UAVs can also support the ground nodes in the routing process. They UAVs can be partially shifted to be located between out-of-range nodes to allow them transfer their data.
- **Flying Road Side Unit:** Enabled with Dedicated Short Range Communications (DSRC), the UAV will be used as a flying road side unit that can fly to a specific location to act as a V2X RSU (e.g., extend communication link at corners) or to broadcast useful information, such as traffic situation in the surrounding area, and suggest alternative detours. In this case, UAV path planning approaches can be developed to determine the locations of UAVs through which it has to pass by to collect the data from ground vehicles. Data routing approaches and clustering algorithms can be combined together to find how data can be efficiently collected in traffic network using UAVs.
- **Flying Police Eye:** The UAV will provides the police agent with a top view video streaming to better assess the traffic around and easily detect specific traffic violations. If the UAV is not directly communicating with police vehicle then, data routing is required through the UAV-assisted VANET.

Joint optimization of the data routing procedure in UAV-assisted VANET using traditional routing protocols and UAV path and/or location adjustments is a new trend research direction that has to be investigated in academia and industry. The mobility of UAVs represents, at the same time, an advantage and a challenge that has to be well studied. Mechanisms enabling efficient coordination and routing among UAVs and close ground vehicles need to be developed while taking into account the specific characteristics of 3D mobile environment, the energy limitation, and the application objective. In the following section, we investigate an example of data routing in UAV-assisted VANET involving the adjustment of some UAVs' locations to ensure an efficient data routing for delay-tolerant application.

IV. Selected Example and Simulation Results

In this section, we provide an example of data routing involving location adjustments of UAVs for delay-tolerant applications. In the framework, we consider a multiple-UAV network where each UAV is executing a certain task related to a primary application. Some of the UAVs are selected to participate to a secondary task, i.e., data routing, in order to transfer data from a source to a destination. In case of the absence of direct communication link, some UAVs are shifted to ensure seamless transmission. The participating UAVs are chosen according to their battery state, mobility range tolerated by the primary application, and the channel quality. The objective is to minimize the data transfer time, denoted by T^{tr}, from the source to the destination which is the sum of the total transmission time in addition to the mobility time denoted by T_n^f. To this end, the following mixed integer nonlinear programming problem is formulated:

$$\underset{\boldsymbol{\epsilon},\boldsymbol{\pi},\boldsymbol{X}^f}{\text{minimize}} \quad T^{tr} = \max_n \pi_n T_n^f + \sum_{n=1}^{N} \epsilon_{sn} T_{sn}^c(X_s, X_n^f) + \sum_{n=1}^{N} \sum_{\substack{m=1 \\ m \neq n}}^{N} \epsilon_{nm} T_{nm}^c(X_n^f, X_m^f) + \sum_{n=1}^{N} \epsilon_{nd} T_{nd}^c(X_n^f, X_d) \quad (5)$$

subject to

$$\sum_{\substack{m=1 \\ m=s}}^{N} \epsilon_{mn} E_{nm}^r + \sum_{\substack{m=1 \\ m=d}}^{N} \epsilon_{nm} (E_{nm}^c) + \pi_n E_n^f \leq \bar{E}_n, \quad \forall n = 1, \ldots, N, \quad (6)$$

$$\pi_n D_n(X_n^0, X_n^f) \leq \bar{D}_n, \quad \forall n = 1, \ldots, N, \quad (7)$$

$$\sum_{m=1}^{N} \epsilon_{mn} = \sum_{n=1}^{N} \epsilon_{nm}, \quad \forall n = 1, \ldots, N, \quad (8)$$

$$\sum_{m=1}^{N} \epsilon_{nm} \leq 1, \quad \forall n = 1, \ldots, N, \quad (9)$$

$$\epsilon_{nm} + \epsilon_{mn} \leq 1, \quad \forall n, m \in \{1, \ldots, N\}, \quad (10)$$

$$\sum_{n=1}^{N} \epsilon_{sn} = 1, \text{ and } \sum_{n=1}^{N} \epsilon_{nd} = 1, \quad (11)$$

$$\epsilon_{nm} \leq \Phi_{nm}, \quad \forall n, m \in \{1, \ldots, N\}, \quad (12)$$

The decision variables of the optimization problem given in (5)-(6) are the binary matrix $\boldsymbol{\epsilon}$, the binary vector $\boldsymbol{\pi}$, and the matrix $\boldsymbol{X^f}$. The entries $\epsilon_{nm}$of the matrix $\boldsymbol{\epsilon}$ denote the states of the links between UAV n and UAV $m, \forall n, m \in \{1, \ldots, N\}$ where N is the number of UAVs in the network. The parameters $\epsilon_{sn}$and ϵ_{nd} indicate the states of the link between the source and a UAV n and the UAV n and the destination, respectively. The entries of $\boldsymbol{\pi}$ are binary variables indicating whether the UAV n is moving from an initial location to another to support the data transfer. If yes, $\pi_n = 1$. Finally, the entries of the matrix $\boldsymbol{X^f}$ correspond to the coordinates of the new locations of N UAVs. Note that $X_n^f = X_n^0$ where X_n^0 are the initial locations of the UAV n. In (5), the first term indicates that the data transmission begins when all the UAVs have reached their new locations. The other terms correspond to the transmission time over the selected path where T_c^{xy} is the communication time needed to send the message from node x to node y. The communication time depends on the size of the message and the achieved throughput per each selected link.

Constraint (6) indicates that the total energy consumed by the UAV during the shifting (E_n^f), the data transmission (E_{nm}^c), and the reception (E_{nm}^r) has to be less than the available energy in its battery allocated to the secondary task. In other words, we assume that, for each UAV, the operator assigns a certain amount of energy for the secondary task that we denote by $\bar{E}_n$. Constraint (7) indicates that, a UAV cannot be shifted with a distance D_n higher than $\bar{D}_n$ in order to guarantee the safe operation of the primary task. The data flow conservation is guaranteed by constraint (8), which forces a UAV that received a data to forward it to other UAVs. Constraint (9) ensures that a data is transmitted to only one UAV and, with constraint (10), cyclic transmission within a single link is disabled. Cyclic data routing over the whole network is avoided by jointly imposing the constraints (8)-(10). Hence, a UAV will not receive the message twice or more during the routing. The equality constraints in (11) force the source to transmit the data and the destination to receive it. Finally, constraint (12) indicates that the data transfer can only be possible over seamless links defined by the binary parameter $\Phi_{nm} = 1$.

The formulated optimization problem given in (5)-(12) is a mixed integer non-linear programming (MINLP) problem. It is difficult and complex to reach the optimal solution of these non-convex problems due to the existence of combinatorial decision variables. Therefore, to solve it, sub-optimal deterministic or meta-heuristic algorithms can be implemented. In the following simulation results, we employ an exploratory search strategy to solve the MINLP problem. Inspired from the Hooke-Jeeves search method applied for multimodal functions, the proposed algorithm tries to find the best directions and distances according to which the UAVs will move in order to establish seamless links for data routing. The objective is to find the best shift combinations for the UAVs that do not affect their energy budgets and do not lead to a high delay. Notice that, for fixed UAV positions, the optimization problem is transformed to an integer linear programming problem that can be optimally solved using CPLEX optimizer.

Four different scenarios are studied in Figure (2). The UAVs' speed are set to 10 m/s and the size of the transmitted message is 10 kilobyte. Figures 2(a), 2(b), and 2(c) investigates the same network topology but for different energy budget distributions. Figure 2(d) illustrates another network topology. Figure 2(a) assumes the absence of energy constraints of all UAVs. The obtained routing path is directly obtained without the need to shift the UAVs with a total transmission time equal to 0.18 s. In Figure 2(b), UAV 4 and UAV 11 are not able to participate in the routing process due to their limited energy budgets ($\bar{E}_4 = \bar{E}_{11} = 0$). Hence, some direct links are degraded and a shifting operation is needed by some of the remaining UAVs. To this end, the exploratory approach shifts UAV 3 and UAV 2 and accordingly, UAV 5 and UAV 6 to establish direct link between the different nodes. Then, instead of transmitting the message to UAV 7, UAV 5 decides to forward it to UAV 2 and UAV 2 skips UAV 11 and sends the message to UAV 6. The total transmission time is equal to 6.17 seconds due to the shifting operation.

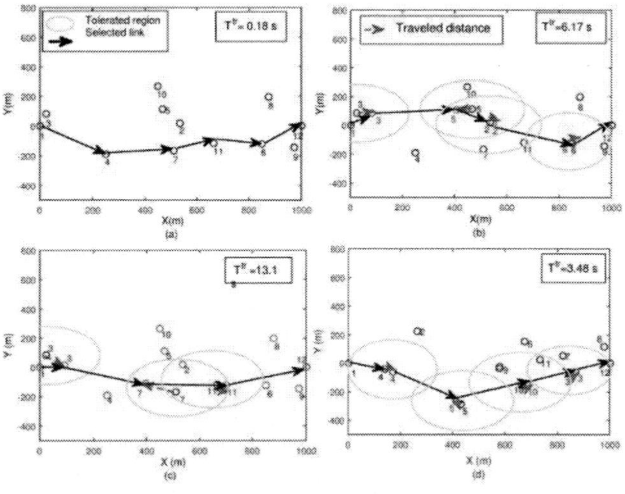

Figure 2. Example of joint location adjustments using an exploratory search strategy (a) scenario 1: no energy limit, (b) scenario 1: $\bar{E}_4 = \bar{E}_{11} = 0$, (c) scenario 1: $\bar{E}_2 = \bar{E}_4 = \bar{E}_5 = \bar{E}_6 = \bar{E}_8 = \bar{E}_8 = \bar{E}_9 = 0$, and (d) scenario 2: no energy limit.

In Figure 2(c), we disable the contributions of all UAVs except UAVs 3, 7, 10, and 11. In this case, the approach decides to shift UAV 7 and UAV 3 in addition to UAV 11. The obtained transmission time is 13.1 seconds due to the high shifted distance (UAV 7).

In Figure 2(d), the initial topology created by the 10 UAVs imposes the execution of a shifting process in order to find a routing path over the FANET. In this scenario, we notice that four UAVs are slightly shifted: UAV 4, UAV 5, UAV 10, and UAV 3. These minor location adjustments result in a low total transmission time of 3.48 seconds.

V. CONCLUSION

In this paper, we investigated the data routing process in UAV-assisted VANETs. We started by highlighting the advantages of using relays in ITS and the challenges that have to be addressed to ensure efficient data routing. Afterwards, we surveyed the data routing techniques that can be implemented in such scenarios by describing some interesting use-cases. Finally, we highlighted the need to proceed with location and/or path adjustments of UAVs in order to obtain direct communication links and seamless routing process. A particular scenario involving UAV location adjustment is finally investigated. The joint optimization of the routing process along with the location and path modifications remain a challenging research direction especially for delay-tolerant applications.

ACKNOWLEDGMENT

This work was made possible by NPRP grants # 8-2459-1-482 and # 9-257-1-056 from the Qatar National Research Fund (A member of The Qatar Foundation). The statements made herein are solely the responsibility of the authors.

REFERENCES

[1] H. Menouar et al., "UAV-enabled intelligent transportation systems for the smart city: Applications and challenges," IEEE Commun. Mag., vol. 55, no. 3, pp. 22–28, Mar. 2017.

[2] E. Uhlemann, "Connected-vehicles applications are emerging [connected vehicles]," IEEE Veh. Technol. Mag., vol. 11, no. 1, pp. 25–96, Mar. 2016.

[3] A. M. Vegni and V. Loscr, "A survey on vehicular social networks," IEEE Commun. Surveys Tuts., vol. 17, no. 4, pp. 2397–2419, Fourth quarter 2015.

[4] K. Abboud, H. A. Omar, and W. Zhuang, "Interworking of DSRC and cellular network technologies for V2X communications: A survey," IEEE Trans. Veh. Technol., vol. 65, no. 12, pp. 9457–9470, Dec. 2016.

[5] C. Bila, F. Sivrikaya, M. A. Khan, and S. Albayrak, "Vehicles of the future: A survey of research on safety issues," IEEE Trans. on Intell. Transp. Syst., vol. 18, no. 5, pp. 1046–1065, May 2017.

[6] A. Awang, K. Husain, N. Kamel, and S. Aissa, "Routing in vehicular ad-hoc networks: A survey on single- and cross-layer design techniques, and perspectives," in IEEE Access, vol. 5, pp. 9497-9517, Apr. 2017.

[7] S. Hayat, E. Yanmaz, and R. Muzaffar, "Survey on unmanned aerial vehicle networks for civil applications: A communications viewpoint," IEEE Commun. Surveys Tuts., vol. 18, no. 4, pp. 2624–2661, Fourth quarter 2016.

[8] A. Al-Hourani, S. Kandeepan, and S. Lardner, "Optimal LAP altitude for maximum coverage," IEEE Wireless Communications Letters, vol. 3, no. 6, pp. 569–572, Dec. 2014.

[9] H. Ghazzai, H. Menouar, and A. Kadri, "On the placement of UAV docking stations for future intelligent transportation systems," in IEEE Veh. Technol. Conf. (VTC-spring workshops), Sydney, Australia, Jun. 2017.

[10] S. Kaur and M. Talwar, "Routing strategies in flying ad-hoc networks," J. Netw. Commun. Emergi. Technol., vol. 6, no. 3, pp. 59–62, Mar. 2016.

[11] Md. H. Tareque, Md. S. Hossain, and M. Atiquzzaman, "On the routing in flying ad hoc networks," in Federated Conf. Comput. Sci. Inf. Syst., Gdansk, Poland, Sept. 2016.

[12] S. Rosati, K. Kruelecki, G. Heitz, D. Floreano, and B. Rimoldi, "Dynamic routing for flying ad hoc networks," IEEE Trans. Veh. Technol., vol. 65, no. 3, pp. 1690–1700, Mar. 2016.

[13] A. I. Alshbatat and L. Dong, "Cross layer design for mobile ad-hoc unmanned aerial vehicle communication networks," in International Conf. Netw., Sens. and Control (ICNSC 2010), Chicago, IL, USA, Apr. 2010.

[14] M. Benzaid, P. Minet, and K. A. Agha, "Analysis and simulation of fast-OLSR," in IEEE Veh. Technol. Conf. (VTC-spring 2003), Jeju, South Korea, Apr. 2003.

[15] J.-D. M. Me Biomo, T. Kunz, M. St-Hilaire, and Y. Zhou, "Unmanned aerial ad hoc networks: Simulation-based evaluation of entity mobility models' impact on routing performance," Aerospace, vol. 2, no. 3, pp. 392–422, Jun. 2015.

[16] J. D. M. M. Biomo, T. Kunz, and M. St-Hilaire, "Routing in unmanned aerial ad hoc networks: A recovery strategy for greedy geographic forwarding failure," in IEEE Wireless Commun. and Netw. Conf. (WCNC 2014), Istanbul, Turkey, Apr. 2014.

An ITS-based Architecture for Opportunistic Networking

Rodrigo Silva[1,3]
IMT Atlantique[1]
Rennes, France
Email: rodrigo.silva@imt-atlantique.fr

Jean-Marie Bonnin[1,2,3,4]
INRIA[2]/IRISA[3]
Rennes, France
Email: jean-marie.bonnin@irisa.fr

Thierry Ernst[4]
YoGoKo[4]
Paris, France
Email: thierry.ernst@yogoko.fr

***Abstract*—For a smarter use of transportation systems, vehicles need to increase their environment awareness. This could be achieved by enabling vehicles to communicate with their environment. Once vehicles become connected, an ecosystem of applications and services could be developed around them, enabling the information exchange with other connected devices and contributing for a Cooperative Intelligent Transportation Systems (C-ITS). The environment of connected and cooperative vehicles is characterized by its heterogeneity, i.e., there are a wide variety of applications, a variety of users with different communication preferences. Moreover, countries may have specifics regulations. A single access technology to connect all these heterogeneity is impossible. For ubiquitous connectivity it is necessary to use existing wireless technologies (e.g., vehicular WiFi (ITS-G5, DSRC), urban WiFi, 802.15.4, and cellular). In such heterogeneous network environment, applications and services cannot take into account all technology particularities. It is necessary a communication architecture that hides the underlying differences of access networks from applications, providing seamless communication independently of the access technology. Based on the ITS architecture proposed by International Organization for Standardization (ISO) and European Telecommunications Standards Institute (ETSI), we propose a Decision Maker (DM) architecture that is capable to manage requirements and preferences from different actors (e.g., applications, users, administrators and regulators), it takes into account the short-term prevision about the network environment, it considers the context information (e.g., vehicle speed, battery level). And it also takes into account the route conditions between two communicating devices in order to make proactive decisions.**

***Keywords*—ISO TC 204; ETSI TC ITS; ITS station communication architecture; C-ITS; decision making.**

I. INTRODUCTION

The number of connected devices is growing fast around the world. These objects are components of a network known as the Internet of Things (IoT), where each object has the possibility to exchange data with others. This scenario enables the development of smart cities, where vehicles are supposed to be one of the communicating objects. According to Gartner research company, connected cars would be a major element of the IoT, representing 20% of all IoT devices [1].

For a smarter use of transportation systems, vehicles need to increase their environment awareness. This could be achieved by enabling vehicles to communicate with their environment. The connection could be local between nearby devices or global, i.e., connection over the Internet. Once vehicles become connected, an ecosystem of applications and services can be developed around them. Nowadays, we are connected to Internet through our computers and smartphones. In the future, the vehicles will be directly connected too, supporting a variety of applications (like smartphones do). For example, vehicles could connect to the Internet to enhance driver and passenger experience, e.g., improving the navigation and offering on-board Internet connectivity. Vehicles can to connect and exchange information with other devices in a smart city environment. In this context, users, devices and vehicles need to be connected anywhere, anytime with anything. Such connections will enable the information exchange between vehicles and their environment for a Cooperative Intelligent Transportation Systems (C-ITS).

However, a single access technology to connect all these heterogeneity of services and devices is impractical or even impossible. For ubiquitous connectivity it is necessary to use existing wireless technologies, such as vehicular WiFi (ITS-G5, DSRC), urban WiFi, 802.15.4, WiMAX, cellular (3G, 4G, and 5G under preparation) [2]–[4]. Each of these networks has specific characteristics in terms of bandwidth, data rate, security and others. Due to this network heterogeneity and its complementary characteristics, more connectivity opportunities are available. Mobile devices equipped with multiple communication capabilities could use multiple access technologies simultaneously in order to maximize flows satisfaction (e.g., to maximize communication bandwidth, to reduce latency, and others) and to satisfy communication requirements (e.g., security, monetary cost, traffic load balancing among available networks, and others).

The environment of connected and cooperative vehicles is characterized by its heterogeneity. For example, there are a wide variety of applications, each one with specific requirements. There are a variety of users with different preferences. Countries could have specifics regulations. There are a variety of access technologies, each one with specific characteristics in terms of bandwidth, data rate, security and others. Moreover, vehicles can move at high speed and frequently change its network environment.

In such heterogeneous and dynamic network environment, applications and services cannot take into account all technology particularities, unless they explicitly need it. The communication architecture has to hide the underlying differences of access networks from applications, providing seamless communication independently of the access technology. It should be capable to handle multiple access technologies simultaneously while select the most appropriate access network for each flow. Such an architecture should choose the path, i.e., the route between two communicating nodes that best meets the communication requirements (e.g., a local connection between nearby devices or a global connection over the Internet). Moreover, in such dynamic environment the communication architecture should perform proactive decisions taking into account the short-term prevision about the network availability.

Based on some research work and also based on the ITS architecture proposed by International Organization for Standardization (ISO) and European Telecommunications Standards Institute (ETSI), we propose a Decision Maker (DM) architecture. Such architecture is capable to manage requirements and preferences from different actors (e.g., applications, users, administrators and regulators), it takes into account the short-term prevision about the network environment, it considers the context information (e.g., vehicle speed, battery level). And it also takes into account the route conditions between two communicating devices in order to make proactive decisions. The proposed DM architecture is developed in an ISO/ETSI standard compliant way.

The remainder of this paper is organized as follows. Section II overviews the reference ITS station communication architecture proposed by ISO and ETSI. Section III reviews some related work. The proposed DM architecture is described in Section IV. And Section V concludes the paper and proposes future directions.

II. The Reference ITS Station Communication Architecture

In order to establish an harmonized communication-centric architecture for ITS, ISO and ETSI have proposed a reference ITS communication architecture supported by nodes called ITS Stations (ITS-S), where each ITS-S (e.g., vehicles) can handle its communication [5]. Such architecture is shown on Figure 1.

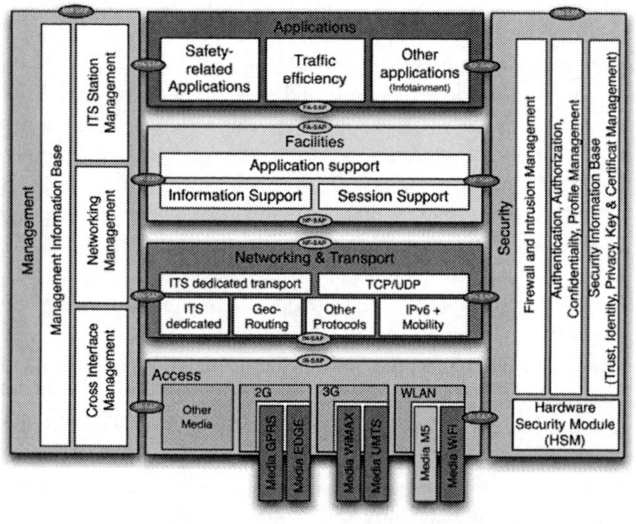

Figure 1. The reference ITS station communication architecture.

The concept of the ITS-S communication architecture is to abstract applications from both the access technologies and the networks that transport the information between communicating nodes. Therefore, applications are not limited to a single access technology, but they could take advantage from all available technologies. While the lower layers can be independently managed, without impacting applications.

In such architecture, two cross layers entities, i.e., "ITS Station Management" and "ITS Station Security" are responsible to station management functionalities and to provide security and privacy services, respectively. Since applications are developed regardless to communication networks, "ITS Station Management" entity is responsible to choose the best network interface for each application. In order to manage different process in the ITS-S, such cross layers entities communicate with the horizontal layers: "ITS Station Access Technologies" layer that is responsible for media access control and provides data transmission through different access technologies, such as vehicular WiFi (ITS-G5, DSRC), urban WiFi, 802.15.4, WiMAX, and cellular (3G, 4G, and 5G under preparation); "ITS Station Networking & Transport" layer, which is responsible to execute operations like packet routing, path establishment, path monitoring and Internet Protocol (IP) mobility; "ITS Station Facilities" layer that provides applications, information and communication supports (e.g., encode/decode message support, time-stamping and geo-stamping) and "ITS Station Application" layer that provides Human-Machine Interface (HMI).

Network Mobility Basic Support Protocol (NEMO) [6] has been chosen by several standardization bodies for IP-based mobility management, including ISO and ETSI. NEMO allows a Mobile Router (MR) to manage the IP mobility for all mobile network attached to it. The MR maintains a bi-directional tunnel (protected by IPsec) to a server in the cloud referred to as the Home Agent (HA), as shown on Figure 2. For the mobile network, it is allocated an IPv6 prefix identifying the mobile network in the IP addressing topology as permanently attached to the HA. Based on this prefix, the MR assigns unchangeable addresses to its attached nodes called Mobile Network Nodes (MNN). When a new network is available, MR generates a new auto configured IP address (Care-of-address (CoA)) within the new visited network and notifies them to the HA. Only the MR and the HA are aware of the network change, since MNNs remain connected to the MR through their permanent IP address.

MRs can be equipped with multiple communication interfaces. Multiple Care of Addresses Registration (MCoA) [7] is used to managed these communication interfaces simultaneously, as illustrated on Figure 2. MCoA enables the registration of several CoAs for a single MR. In this case, the MR could establish multiple tunnels through each of its communication interfaces and the HA.

The possibility of having multiple applications in an ITS-S simultaneously competing for communication resources leads to the need for a controlled access to these resources. In such a control, requirements and objectives presented by

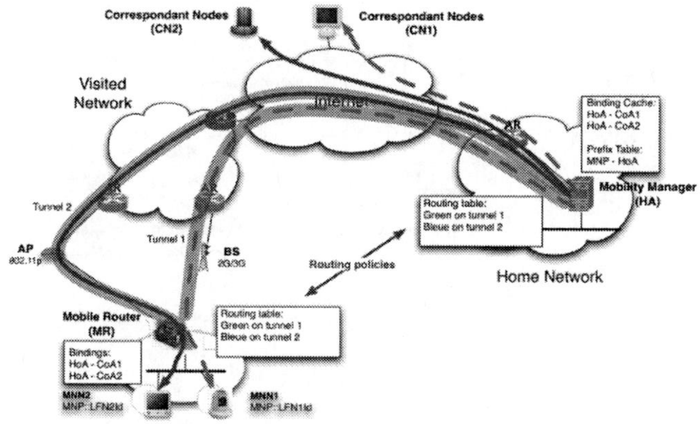

Figure 2. NEMO and MCoA.

application, user preferences, set of rules (e.g., regulations, network operator policies, etc.) and communication protocols' status are used by the ITS-S Management Entity (SME), from "ITS-S Management" cross layer, to select the best suited communication profile and path per communication source. The determination of the path implies the selection of the communication interface, the logical node in the access network to which the ITS station is locally attached (ingress anchor node) and the intermediary nodes in the network used to reach the destination node (egress anchor node). Aware about paths characteristics, the SME can choose the path that best meets the communication requirements (e.g., a local connection between nearby devices or a global connection over the Internet). The methods to determine the most appropriate path and to perform flow-interface mapping is implementation specific as it could be a competitive factor between stakeholders. It is thus not specified in the ISO standards.

III. Related work

Few researches have worked on the development of a DM architecture that consider the use of multiple access technologies simultaneously and routing flow per flow, i.e., spreading flows among different communication interfaces.

Authors of [8] proposed a modular architecture for multi-homed mobile terminals. In such architecture, a middleware interacts with "higher-layers" and "low layers". The "high layers" gather the user and the administrator preferences, handle the applications' requirements, and detect the current terminal capabilities. The "low layers" detects the available networks and provides real-time information about the interfaces and access networks capabilities as well as it handles the selection execution process, i.e., it maps the application's flows on the preferred access network. It does not consider the path condition of a given flow between sender and destination nodes. And it does not consider the near future of the network environment, i.e., the short-term prevision about the availability of networks.

Paper [9] proposes a context-aware management solution to maximize the satisfaction of the applications while respecting the stakeholders policy rules. The proposed framework collect and combining policies from stakeholders (e.g., user, administrators and applications). Based on such policies and context information, it evaluates the network that better match the communication requirements. Once the best network is chose, the flow routing is enforced on the device using NEMO and MCoA. Such architecture does not consider the path condition experienced by a flow or the near future of the network environment.

Paper [10] proposes a framework for supporting network continuity in vehicular IPv6 communications. Such framework follows the ISO/ETSI guidelines for the development of cooperative ITS systems and is based on standardized technologies, such as NEMO protocol to provide an integral management of IPv6. However, it considers cooperation between mobile devices and networks based on the 802.21 standard (Media Independent Handover (MIH)), i.e., it considers that all networks support the specific functionalities from 802.21 standard [11].

IV. The ITS-based DM Architecture

This section describes the modular DM architecture for opportunistic networking in heterogeneous access network environment. The proposed architecture is based on the previously described ITS-S communication architecture and designed to meet the main challenges for communication profile and path selection in C-ITS environment.

A. Expected properties

As described by [12], the environment of connected and cooperative vehicles is characterized by a large heterogeneity. There are a wide variety of applications with different communication requirements. There are different wireless access technologies each one with specific characteristics in terms of bandwidth, data rate, security and others. In such an environment, the process to select the best suited communication profile and path for each data flow presents some challenges.

Different actors are able to present their requirements, preferences, constraints and policies in the decision making process. For example, applications can request a specific bandwidth, data rate or security level. Users can present their preferences, e.g., defining a priority or security level for a given message. Industrial and mobility service providers (i.e., operators) can present their policies, such as network constraints and particular billing procedures. Moreover, these wide variety of objectives could be contradictory. The DM architecture should be capable of managing these multiple objectives simultaneously.

The DM architecture should be able to monitor a variety of information in order to enable more accurate solutions in the decision making process. One essential piece of information to be monitored is the wireless networks availability as well as the performance of the networks in use. Moreover, it is necessary to monitor flows and their characteristics (e.g., used bandwidth, flow status).

Besides network monitoring, other significant parameters could be monitored. Vehicles would be able to take information from their environment, as vehicle's battery level, geographical position (e.g., GPS) or vehicle's speed in order to adjust the decision's strategies. For example, a power consuming network interface could be deactivated if the vehicle's battery level is under a certain threshold. Or a WiFi network could be privileged if the vehicle is stationary, while a cellular network could be preferred if the vehicle is moving.

The DM architecture should be capable of handling communication profile and communication path for each flow. A data flow is defined by ISO as an identifiable sequence of packets [13]. And packets are dependent upon applied protocols, links and nodes characteristics. For example, packets sent over different communication paths (routes) to the same destination node experience distinct network conditions/performances. Such distinct experiences are consequence of the applied protocol stacks (communication profile) and the specific characteristics of the traversed path (e.g., delay, throughput, security level, etc.). Therefore, on the Flow-Interface mapping process, it is not enough to indicate only what access network a given flow should use. In addition, according to flow requirements and paths characteristics it is necessary to determine the communication profile and path for each flow.

Moreover, due the vehicle's high speed the networks availability could change rapidly. In such highly dynamic mobility the decision making process should take into account the short-term prevision about the network environment condition. If the DM is aware about the near future of the network environment it can perform proactive and fine-grained decision. For example, it can decide to increase the data buffer for a given video streaming, if the vehicle is going to cross a wireless dead zone. Or, an on-board application could decide to delay a data transmission if it knows that a better network will soon be available.

The short-term prevision can be obtained in different ways. It can be obtained by cooperation with networks, e.g., using the IEEE 802.21 standard if the network support such protocol. The vehicle can store network information from a previous traversed route, e.g., for an user who uses the same route every day, the database could stores information about network conditions in such route. Or, the short-term information can be obtained by cooperation with neighbors vehicles. For example, two vehicles in opposite directions could exchange information about access points in their upcoming route. For this purpose, a vehicle stores the position of each access point in its traversed route, and give them to another passing-by vehicle.

B. Architecture design

To achieve the expected properties, we propose a modular DM architecture based on the ISO/ETSI standards. Figure 3 shows such proposed DM architecture.

For a better understanding, we split the DM architecture in five main parts, which are described below.

1) Requirement gathering: As mentioned before, different actors are able to present theirs requirements in the decision making process. In our proposed architecture we consider four main actors. *Applications* - it could request a specific

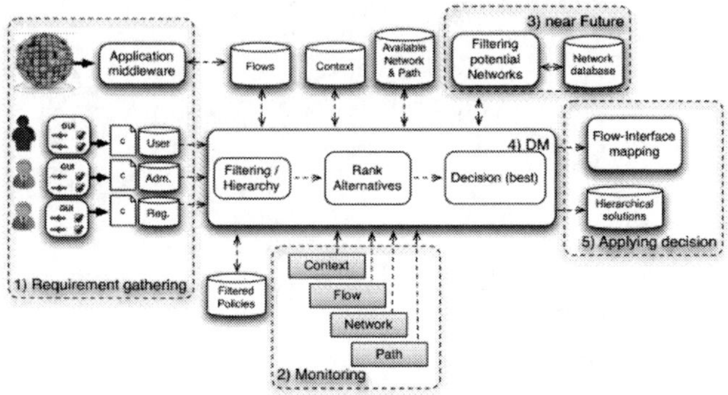

Figure 3. Proposed Decision Maker Architecture.

bandwidth, data rate, security level and more. A middleware enables different applications to send their requirements to DM. *Users* - they can present their preferences through a Graphical User Interface (GUI), e.g., defining priority or security level for a given message. *Administrators*, i.e., industrial and mobility service providers - they could present their policies, such as network constraints and particular billing procedures. And *regulator* bodies - each country or region could define some specific rules, such as the prohibition of certain frequency ranges in certain areas. The requirements from all actors are stored in decision maker's databases and used by the DM to choose the communication interface that better matches the actors requirements.

2) Monitoring modules: We defined four monitoring modules. *Network monitoring module* - in this process, the network monitoring module listens to the wireless interfaces and informs DM about the available wireless networks and their performances. Such monitoring module should be able to monitor network information even if no specific monitoring functionality, such as 802.21 [11], is implemented on the network side. *Context monitoring module* - this module is responsible for vehicle surrounding monitoring. It is responsible to monitor information like location of the neighboring vehicles, traffic jam, vehicle's speed, and others. These information are part of the Local Dynamic Map (LDM) functionalities, i.e., the conceptual data store located within an ITS-S as outlined in [14]. Therefore, we aim to rely this monitoring module on such conceptual data store. *Flow monitoring module* - this module should inform whether a flow is alive or not and evaluate flows' performance, like the currently used bandwidth, the currently latency, etc. *Path monitoring module* - this module is responsible to obtain various information (e.g., throughput, security level, latency, etc.) about the controllable end points where packets will be routed and to keep track of all the candidate and available paths.

3) Near Future: In order to take into account the short-term prevision about the network environment, we propose a network database that store the historical information about the access networks (e.g., network performance and access point location) and a filtering entity that is responsible to analyze such network database and, based on the context information of the vehicle (e.g., movement speed), to choose the potential networks to be considered in the decision making process.

4) Decision making process: The decision making process is responsible to take into account the application's requirements, user profiles, administrative rules (regulation and policies) as well as different monitored information in order to manage flows and paths. The decision making process is detailed in section IV-C.

5) Applying decision: In the applying decision process, the policies and information produced by the decision making process are applied in the system. In this process, the decision maker could interact with controlled entities in all layers of the ITS station communication architecture. Once the best access network and path is selected, i.e., the path and access network that better match the communication requirements, the DM request the "Flow-Interface mapping" module to enforce the flow routing decision. To enforce the decision's polices at the network layer in an IP-based environment, we are considering NEMO and MCoA. These protocols allow mobile routers to manage multiple access technologies simultaneously and to improve path and flow management.

Since the decision making process take into account the short-term prevision about the network environment, proactive decisions are enforced in order to maintain flows always best connected. However, unexpected changes can occur in a wireless environment (e.g., a given access network can drops). In order to adapt to the network conditions in real time, the DM maintain an hierarchical solution database with all sub-optimal solutions for each flow. This database is used by the "Flow-Interface mapping" module in case of emergency, i.e., when the best network solution drops unexpectedly and until the DM finds another better solution.

C. Decision Making Process

As mentioned before, the decision making process takes into account the application's requirements, user profiles, administrative rules as well as information from a variety of monitoring modules in order to manage flows and paths. We split our decision making process in three modules, as shown on Figure 3. Below we describe each one of these modules:

Hierarchy/Filtering: This module is responsible to receive and manage requirements, preferences, and policies from different actors. Since actors may have their own specific preferences and requirements, we need to "filter" (in Computer Science acceptation) the various values defined for the same parameter. Moreover, it is necessary to define a priority order between actors in order to manage contradictory objectives. For example, if the administrator sets a forbidden network for a user, and the user set the same access network to preferred, then it is necessary to define who has the priority. The output of such module is a list of filtered and hierarchical requirements.

Rank Alternatives: This module is responsible to find all alternatives for flow-interface mapping. It is a first filter to avoid forbidden networks or networks that do not match with flows' requirements. Such module receives the coherent list of requirements from "Hierarchy/Filtering" module, network information (e.g., networks availability and networks performance), and context information in order to find the potential solutions. The output of this module is a list of all potential solutions for each flow.

Decision Algorithm: This module receives the list of all potential solutions created in the "Rank Alternatives" module and apply decision making algorithm in order to evaluate the matching degree of communication requirements with networks and path characteristics. An utility function calculates a score, representing the matching degree for each solution. Higher the score, better is the solution. The solutions are sorted by descending order of score and stored on the hierarchical solution database. Such database is used by the enforcement module in case of emergency, i.e., when the best network drops unexpectedly, the "Flow-Interface mapping" module redirect the flow through the first available sub-optimal network while the DM finds a new better solution.

As described by [12], several decision making algorithms have been used in the network selection process. For example, the ones based on the game theory, the ones based on Multi-Objective Optimization (MOO) and the algorithm that uses Multi-Attribute Decision Making (MADM) techniques. The most used are the MADM methods (e.g., Simple Additive Weighting (SAW), Technique for Order Preference by Similarity to Ideal Solution (TOPSIS) and Analytic Hierarchy Process (AHP)). Despite the MADM techniques present advantage as relative low computation complexity, this approach has some issues. For example, it is very difficult to choose the best weight for each attribute. Moreover, MADM algorithms could present ranking abnormality, i.e., change in one of the parameters of the objective function could determine a very different best solution. The design of a decision making algorithm is outside the scope of this paper. Such topic will be addressed in future works.

D. Integration in the ITS-S communication architecture

The ITS-S communication architecture functionalities could be implemented into a single physical unit, as the practical implementation showed on paper [15], or distributed into several physical units. Once applied to a vehicle, these functionalities could be performed by different modules in the vehicle's electric/electronic architecture.

Moreover, the NEMO environment mainly separate the applications and communications into MNN and MR. Therefore, the five functions described in Section IV-B can also be separated into such nodes. For example, the requirement gathering can be implemented in the MNN, the monitoring modules can be implemented in both MR and MNN, while the near future, the decision making process and applying decision are functions of the MR.

The proposed DM architecture is designed in an ISO/ETSI standard compliant way. Figure 4 shows one way how we can integrate such architecture in the ITS-S communication architecture.

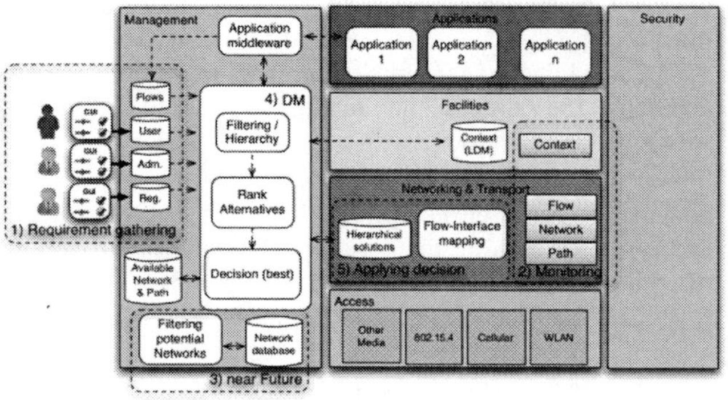

Figure 4. Integration of DM Architecture in the ITS-S communication architecture.

However, the standards give some guidelines to the developers, leaving some room in the way to implement the ITS-S communication architecture.

V. Conclusion and Future Work

In this paper, we proposed a modular decision maker architecture that is capable to choose the best available communication profile and path for each data flow in an heterogeneous and dynamic network environment. The proposed DM architecture is designed in an ISO/ETSI standard compliant way and we show how to integrate it in the ITS-S communication architecture.

Besides the access network selection, the proposed architecture is able to choose the best path for a given flow, i.e., the route between two communicating nodes that best meets the communication requirements (e.g., a local connection between nearby devices or a global connection over the Internet).

Different actors are able to present theirs requirements in the decision making process, e.g., applications, users, network administrators, etc. And this wide variety of objectives could be contradictory. The proposed DM architecture is capable of managing these multiple objectives simultaneously. Moreover, the DM receives information from a variety of monitoring modules (network, context information, path, and flows monitoring modules), that enable fine-grained decisions.

The proposed architecture address the short-term prevision about the network environment. This short-term prevision allows proactive decisions, which is very useful in vehicular environments that are characterized by highly dynamic mobility.

Once the best access network and path is selected for a given flow, the decision's polices are enforced at the network layer using standardized protocols, such as NEMO and MCoA. These protocols allow mobile routers to manage multiple access technologies simultaneously and to improve path and flow management.

We highlight the importance of the DM architecture validation. As future work, we will simulate the proposed architecture using different scenarios and existing decision making algorithms. We will also design and simulate a decision making algorithm capable to take advantage of the entire proposed architecture for smart and fine-grained decisions. Moreover, it is valuable to conduct such architecture in a real test-bed.

References

[1] "Gartner Says By 2020, a Quarter Billion Connected Vehicles Will Enable New In-Vehicle Services and Automated Driving Capabilities," http://www.gartner.com/newsroom/id/2970017, [Accessed: June 2017].

[2] "ETSI ES 202 663 European profile standard for the physical and medium access control layer of Intelligent Transport Systems operating in the 5 GHz frequency band," Tech. Rep., 2010.

[3] "IEEE 802.15.4 – Local and metropolitan area networks– Specific requirements– Part 15.4: Wireless Medium Access Control (MAC) and Physical Layer (PHY) Specifications for Low Rate Wireless Personal Area Networks (WPANs)," Tech. Rep., 2006.

[4] S. Chakrabarti, G. Montenegro, and R. Droms, "IPv6 over Low-Power Wireless Personal Area Network (6LoWPAN) ESC Dispatch Code Points and Guidelines," RFC 8066, 2017.

[5] "ISO 21217 Intelligent Transport Systems — Communications Access for Land Mobiles (CALM) — Architecture," Tech. Rep., 2014.

[6] A. Petrescu, R. Wakikawa, P. Thubert, and V. Devarapalli, "Network Mobility (NEMO) Basic Support Protocol," RFC 3963, 2005.

[7] H. Soliman, G. Tsirtsis, G. Giaretta, K. Kuladinithi, and N. Montavont, "Flow Bindings in Mobile IPv6 and Network Mobility (NEMO) Basic Support," RFC 6089, 2011.

[8] L. Suciu, J. M. Bonnin, K. Guillouard, and B. Stevant, "Towards a highly adaptable user-centric terminal architecture," in *The 7th Intl. Symposium on Wireless Personal Multimedia Communications (WPMC)*, Italy, 2004.

[9] R. Ben Rayana, "A Smart Management Framework for Multihomed Mobile Nodes & Mobile Routers," Doctoral Dissertation, Jan. 2010.

[10] J. Santa, F. Pereniguez-Garcia, F. Bernal, P. Fernandez, R. Marin-Lopez, and A. Skarmeta, "A Framework for Supporting Network Continuity in Vehicular IPv6 Communications," *IEEE Intelligent Transportation Systems Magazine*, vol. 6, no. 1, pp. 17–34, Spring 2014.

[11] "IEEE 802.21 – Standard for Local and metropolitan area networks – Part 21: Media Independent Handover Services," Tech. Rep., 2008.

[12] R. Silva, J.-M. Bonnin, and T. Ernst, "Opportunistic Networking for ITS," in *Intelligent Transport Systems: Past, Present and Future Directions*. Nova publisher, 2017.

[13] "ISO/DIS 24102-6 Intelligent Transport Systems — Communications Access for Land Mobiles (CALM) — ITS station management — Part 6: Path and flow management," Tech. Rep., 2015.

[14] "ETSI EN 302 895 Intelligent Transport Systems (ITS); Vehicular Communications; Basic Set of Applications; Local Dynamic Map (LDM)," Tech. Rep., 2014.

[15] R. Silva, S. Noguchi, T. Ernst, A. De La Fortelle, and W. Godoy Junior, "Standards for Cooperative Intelligent Transportation Systems: A Proof of Concept," in *The Tenth Advanced International Conference on Telecommunications (AICT)*, Jul. 2014.